Environmental Engineering and Science

Environmental Engineering and Science

June Banks

Callisto Reference
www.callistoreference.com

Callisto Reference,
118-35 Queens Blvd., Suite 400,
Forest Hills, NY 11375, USA

Visit us on the World Wide Web at:
www.callistoreference.com

ISBN: 978-1-64116-755-0 (Hardback)

Cataloging-in-Publication Data

Environmental engineering and science / June Banks.
p. cm.
Includes bibliographical references and index.
ISBN 978-1-64116-755-0
1. Environmental sciences. 2. Environmental engineering. 3. Environmental management.
4. Environmental protection. I. Banks, June.
GE105 .E58 2023
363.7--dc23

Table of Contents

Preface

Environmental science consists of the application of scientific and engineering principles to improve and maintain the environment to protect human health, to protect nature's beneficial ecosystems, and to improve environment-related quality of human life. Environmental engineering is an engineering discipline concerned with the application of scientific and engineering principles to improve and maintain the quality of environment. While environmental science focuses on collecting and analyzing data as well as making hypothesis on the basis of this data, environmental engineering involves the application of this data for the production of outcomes which lead to reducing the impacts of development and population growth. The book studies, analyses and explains the most important concepts in environmental engineering and science. The topics included herein are of utmost significance and bound to provide incredible insights to readers. The book is a resource guide for experts as well as students.

Various studies have approached the subject by analyzing it with a single perspective, but the present book provides diverse methodologies and techniques to address this field. This book contains theories and applications needed for understanding the subject from different perspectives. The aim is to keep the readers informed about the progresses in the field; therefore, the contributions were carefully examined to compile novel researches by specialists from across the globe.

Indeed, the job of the editor is the most crucial and challenging in compiling all chapters into a single book. In the end, I would extend my sincere thanks to the chapter authors for their profound work. I am also thankful for the support provided by my family and colleagues during the compilation of this book.

June Banks

Chapter 1

Environment, Ecosystems and Biodiversity

1.1 Environment

Environment is defined as the surroundings or conditions in which a person, animal or plant lives or operates.

Some important definitions of environment are as under:

- Boring: An environment consists of the sum of the stimulation which receives from conception until death. It can be concluded from the above definition that, "Environment comprises of various types of forces such as physical, political, intellectual and economic, social, cultural, moral and emotional".
- Douglas and Holland: The term environment is used to describe all the external forces, influences and conditions, which affect the life, nature, behaviour and the growth, development and maturity of living organisms.

Environmental Study

Environment study deals with the analysis of the processes in air, water, soil, land and organisms which lead to pollute or degrade the environment. It helps us for establishing the standard for safe, clean and healthy natural ecosystem. It also deals with the important issues like safe and clean drinking water, hygienic living conditions, fertility of land and clean and fresh air, healthy food and development.

The environmental studies include not only the study of physical and biological characters of the atmosphere but also the social and cultural factors and the impact of man on environment. Environmental property is the ability to keep up rates of natural resources harvest, pollution creation and non-renewable resource depletion which may be continued indefinitely.

Multidisciplinary Nature of Environmental Studies

The science of Environment studies is a multidisciplinary science because it depends on various disciplines like physics, chemistry, medical science, etc. It is the science of physical phenomena in the environment. It is a multidisciplinary field that draws upon not only its core scientific areas, but also applies knowledge from other non-scientific studies such as law, economic and social science:

- Chemistry: To understand the molecular interactions in the system.
- Physics:
    - To construct the mathematical models of environment.
    - To understand the flux of material and energy interaction.
- Atmospheric Science:
    - It comprises of greenhouse gas phenomena, meteorological studies, air-borne contaminants, sound propagation phenomena related to noise pollution and even light pollution.
    - To examine the phenomenology of the Earth's gaseous outer layer with emphasis upon interrelation to other systems.
- Biology: To describe the effects within the plant and animal kingdom and their diversity.
- Environmental Chemistry:
    - Principal area of study includes soil contamination and water pollution.
    - To study the chemical alterations in the environment.
    - The topic of analysis involves multi-phase transport of chemicals, chemical degradation in the environment and chemical effects upon biota.
- Ecology:
    - These studies could also include predator interactions, endangered species, and impact analysis of proposed land development upon species viability or effects upon populations by environmental contaminants.
    - To analyse the dynamics among an interrelated set of population and some aspects of its environment.
- Mathematics and Computer Science: It will help in environmental modeling and analysis of environment related data.
- Geo-science:
    - In some classification systems, it can also embrace hydrology including oceanography.

  - It includes environmental soil science, environmental geology, volcanic phenomena and evolution of earth's crust.

- Law: It helps in framing of environment related Acts, laws, rules and their monitoring.
- Economics: It deals with the economic aspects of various components of environment.
- Social Science: It helps in dealing with the population and health related issues.

1.1.1 Scope and Importance of Risk and Hazards

The terms hazards and risk are often used interchangeable, however, in terms of risk assessment, these are two very distinct terms.

As defined above, a hazard is any biological, chemical, mechanical or physical agent that is reasonably likely to cause harm or damage to humans or the environment with sufficient exposure or dose.

Risk is defined as the probability that exposure to a hazard will lead to a negative consequence.

1.1.2 Chemical Hazards

Most people automatically associate chemicals with scientists in laboratories, but chemicals are also found in many of the products we use at work and at home. While they have a variety of beneficial uses, chemicals can also be extremely harmful if they are misused.

Here are some examples of commonly used household products which may damage our health or cause a fire or explosion if used incorrectly:

- Cleaning products such as toilet cleaners, disinfectants, mildew remover and chlorine bleach.
- Art supplies, such as paint thinner and pottery glazes.
- Garage supplies, such as parts de-greasers and cleaning solvents.
- Office materials, such as photocopier toner.

Chemicals can be broken down into hazard classes and exhibit both physical and health hazards. It is important to keep in mind, that chemicals can exhibit more than one hazard or combinations of several hazards. Several factors can influence how a chemical will behave and the hazards the chemical presents, including the severity of the response:

- Physical state of the chemical such as solid, liquid and gas.

- Concentration of the chemical.
- Physical processes involved in using the chemical such as grinding, cutting, cooling, heating, etc.
- Chemical processes involved in using the chemical.
- Other processes such as addition of moisture, improper storage, storage in sunlight, refrigeration, etc.

Common Chemical Hazards

Specific types of chemicals have been associated with harmful health effects. Common chemical hazards include:

- Skin irritation, disfiguring burns, eye injury or blindness caused by the corrosive chemical products.
- Toxic by-products, such as vapors and fumes, which are caused by mixing incompatible chemicals.
- Serious burns from flammable solvents that catch on fire.
- Poisoning from accidental swallowing, especially with young children.
- Injury from exploding containers, such as spray cans.

1.1.3 Physical Hazards

Physical hazards are those substances or conditions that threaten our physical safety. Fires, temperature, explosive materials, radiation, noise, spills on floors and unguarded machines are some of the examples of physical hazards.

Physical hazards also include ergonomic hazards which occur when the type of work, body position and the working conditions put strain on our body. This happens when our capacity for work is restricted by the type of work. These instances are very hard to spot since we don't always immediately notice the strain on our body or the harm these hazards cause.

Short-term exposure in a badly designed work may result in muscle fatigue or tiredness, but long-term exposure can result in the serious long-term injuries of the muscular-skeletal system. Inner baking is one of the hardest tasks a woman faces routinely.

1.1.4 Biological Hazards in the Environment

Biological hazards, also known as bio hazards, refer to biological substances that pose a threat to the health of living organisms, primarily that of humans. This can include the medical waste or samples of a microorganism, virus or toxin which affects the human health. It can also include the substances harmful to other animals.

1.2 Concept of an Ecosystem

An ecosystem consists of the biological community that occurs in some local and the physical and chemical factors that make up its non-living or abiotic environment. There are many other examples of ecosystems such as a pond, a forest, an estuary and grass-land.

Ecosystem

A group of organisms interacting among them and with the environment is known as ecosystem. Thus, an ecosystem is the community of different species interacting with one another and with their non-living environment exchanging energy and matter.

Example: Animals cannot synthesize their food directly but depends on the plants either directly or indirectly.

Primary Types of Ecosystems

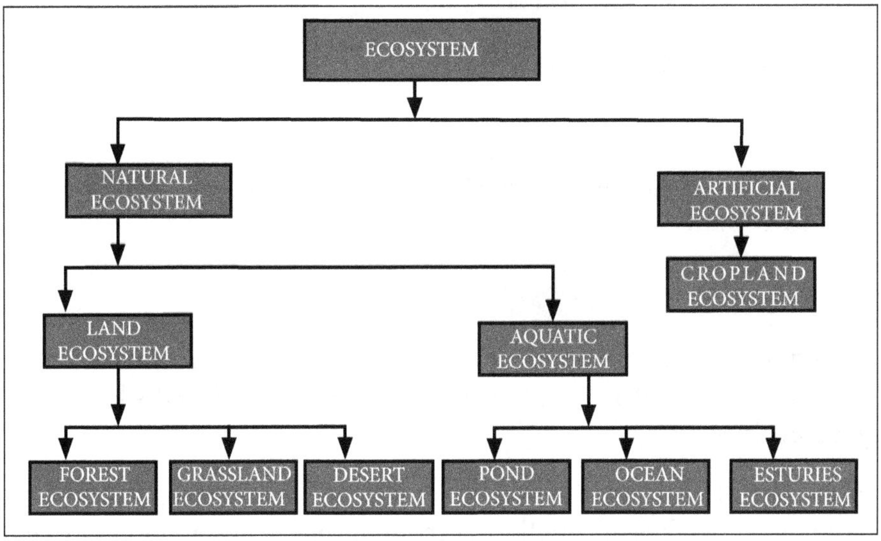

Natural Ecosystems

Natural ecosystems may be terrestrial or aquatic. A natural ecosystem is a biological environment that is found in nature rather than created or altered by man.

Artificial Ecosystems

Humans have modified some ecosystems for their own benefit. These are the artificial ecosystems. They can be terrestrial or aquatic.

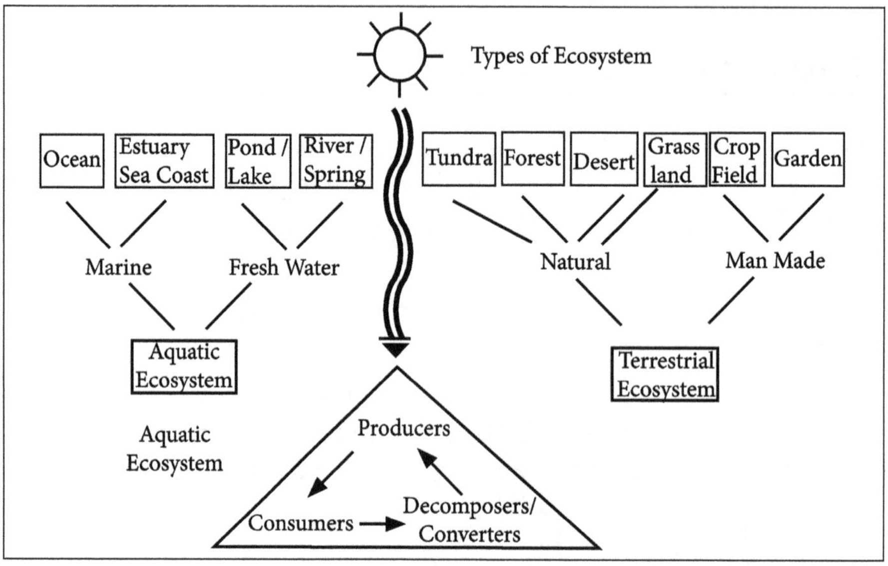

Types of artificial ecosystem.

Types of Natural Ecosystems

There are two main types of natural ecosystems:

- Aquatic natural ecosystem.
- Terrestrial natural ecosystem.

In aquatic ecosystems organisms interact with water. In terrestrial ecosystems organisms interact with land.

Aquatic Ecosystems

Aquatic ecosystems in general cover upto 71% of the earth's surface. As a type, aquatic ecosystems can be classified again into three varieties, defined by the kind of water with which organisms interact:

- Freshwater: This type of ecosystem includes rivers, lakes, streams, ponds and wetlands and makes up smallest percent of the earth's aquatic ecosystems.
- Transitional Communities: These are places where freshwater and saltwater come together, including estuaries and wetlands.
- Marine: More than 70% of the earth is covered by saltwater or marine ecosystems. These include shorelines, coral reefs and open ocean.

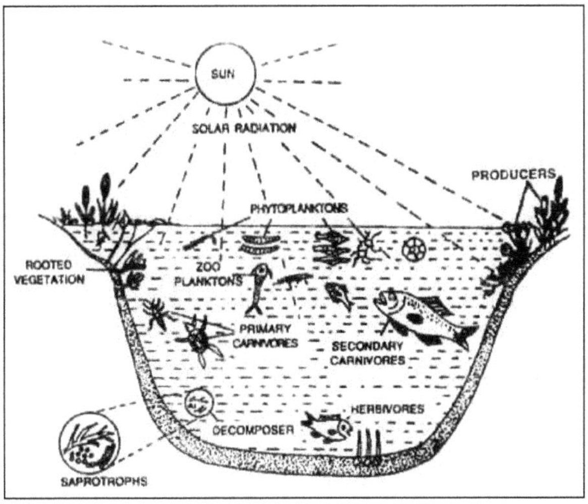

Pond as Eco-system.

Terrestrial Ecosystems

Mountains, forests, deserts and the grasslands are types of terrestrial ecosystems. Terrestrial ecosystems are classified by the type of land or terrestrial area.

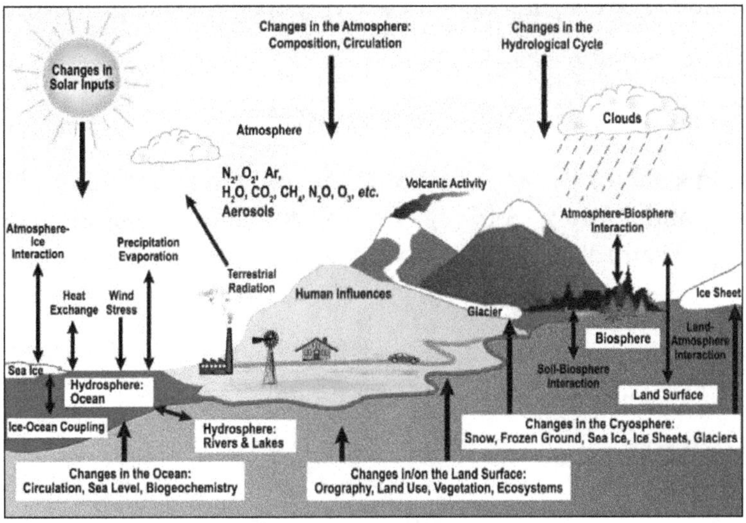

Terrestrial Ecosystems.

- Forest: Forest ecosystems feature dense tree populations and include tropical rain forests.
- Desert: Deserts receive less than 25 cm of rainfall per year.
- Grassland: Grassland ecosystems include tropical savannas, temperate prairies and arctic tundra.
- Mountain: Mountain ecosystems include steep elevation changes between meadows, ravines and peaks.

Balanced Ecosystem

The term ecosystem describes both the living and nonliving components of an area that interact with one another. All the components are interdepartmental in some way with each other. An ecosystem may be aquatic or terrestrial.

In an aquatic ecosystem rocks are needed for shelter and plants provide oxygen for fish. An ecosystem is balanced when natural animals and plants and nonliving components are in harmony i.e. there is nothing to disturb the balance. With increasing pollution, change in migratory patterns and rise of human population, many ecosystems are in danger of losing that harmony.

Advantages

- It is difficult to find a perfectly balanced ecosystem but we can make a model at home and observe how the ecosystem functions. We can observe how different species interact with each other.
- Study the natural cycle of each species.
- Understand the relationship between different species producer, predator and prey.

Creation of a Balanced Ecosystem Model

River silt or clean sand for the aquatic plants to grow. An aquarium tank, at least 1 foot wide, 2 feet deep and 2.5 to 3 feet long. The top of the tank should be detachable and foldable so that we can reach into the aquarium to clean the glass sides and also feed the fish.

A light attached to the top lid to provide artificial light and for warming during cold winters. We can also place the aquarium at a sunny place. Take:

- Aquatic plants.
- Small rocks for the bottom.
- Fish.

First, clean the soil and put it in the bottom of the tank. Arrange the rocks. Then pour a little water into the tank.

Pouring the water directly will disturb the soil bed so pour it over a plastic plate that floats over the surface of the water. This way, we will not disturb the bed or any other plants or fish when we pour water. When the water is at least 1 inch higher than the bed, put plants into the soil. As the water level goes up, the leaves of the plants will also rise up.

After the tank is full with about 4 inches left on top, stop pouring the water. Let the water settle for some time and introduce the fish one by one and close the top.

1.2.1 Structure and Function of an Ecosystem

Structure of an Eco System

The term structure refers to the various components. So the structure of an ecosystem explains the relationship between the biotic and abiotic components.

An ecosystem has two major components:

- Biotic components.
- Abiotic components.

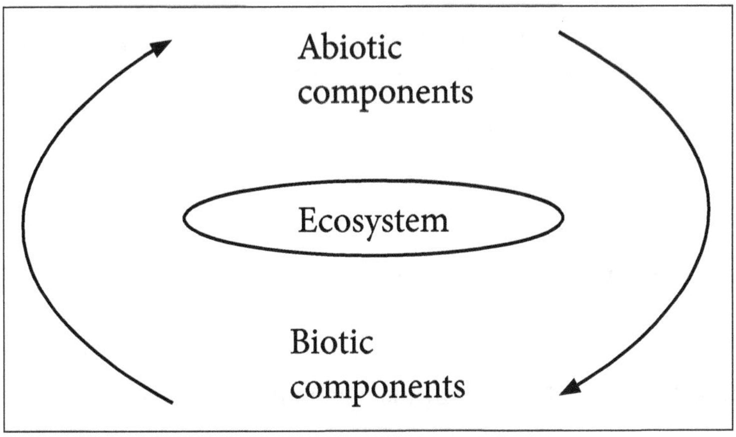

Components of ecosystem.

Biotic Components

The living organisms or living members in an ecosystem collectively form its community called biotic components or biotic community.

Example: Plants, animals and microorganisms.

Classification of Biotic Components

The members of the biotic components of an ecosystem are grouped into three groups based on how they get their food.

- Producers (Plants).
- Consumers (Animals).
- Decomposers (Micro-Organisms).

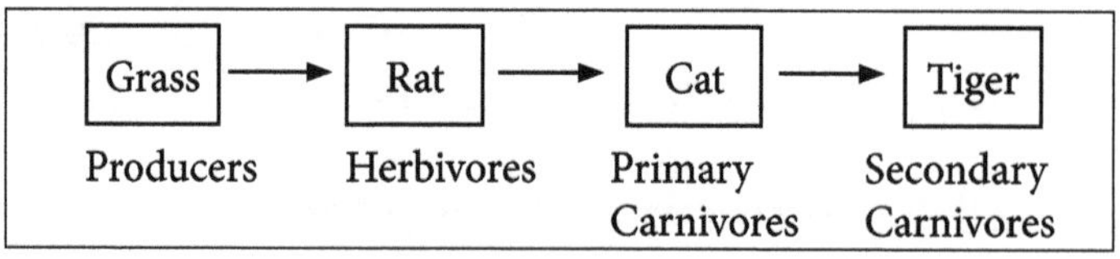

Biotic components of an ecosystem.

Abiotic Components

The non-living components of an ecosystem collectively form a community called abiotic components (or) abiotic community.

Example: Climate, Soil, Water, Air.

Physical Components

They include the climate, energy, raw materials and living space that the biological community needs. They are useful for the growth and maintenance of its member.

Example: Air, Water, Soil, Sunlight, etc.

Chemical Components

They are the sources of essential nutrients.

Example:

- Organic substances: Protein, lipids, carbohydrates, etc.
- Inorganic substances: All micro and macro nutrients and few other elements.

Functions of an Ecosystem

The function of an ecosystem is to allow flow of energy and cycling of nutrients.

Functions of an ecosystem are of three types:

- Primary function: The primary function of all ecosystems is the manufacture of starch.
- Secondary function: The secondary function of all ecosystems is distributing energy in the form of food to all consumers.
- Tertiary function: All living systems die at a particular stage. These dead systems are decomposed to initiate the third function of ecosystems namely "cycling". The functioning of an ecosystem may be understood by studying the following terms:
    - Energy and material flow.

  - Food chains.
  - Food webs.
  - Food pyramids.

1.3 Producers, Consumers and Decomposers

Producers

Producers synthesis their food themselves through the process photosynthesis. Plants are called producers, because they produce their own food. They do this by using light energy from the Sun, carbon dioxide from the air and water from the soil to produce food in the form of glucose/sugar. The process is called photosynthesis.

Example: All green plants, trees.

Consumers

Consumers are organisms, which cannot prepare their own food and depend directly or indirectly on the producers.

- Plant eating species like Insects, rabbit, goal.
- Animals eating species like Fish, Lion, Tiger, etc.

Animals are called consumers, because they cannot make their own food, so they need to consume plants or animals. Humans are also omnivores.

There are 3 groups of consumers:

- Animals that eat only plants.
- Animals that eat only animals.
- Animals that eat both animals and plants.

Classification of Consumers

Consumers are further classified into:

Primary Consumers: Primary consumers are also called herbivores, they directly depends on the plants for their food. So they are called plant eaters.

- Example: Insects, rat, goat.

Secondary Consumers: Secondary consumers are the primary carnivores, they feed on primary consumers. They directly depend on the herbivores for their food.

- Example: Frog, Cat, etc.

Tertiary Consumers: Tertiary consumers are secondary carnivores, they feed on secondary consumers. They directly depend on the primary carnivores for their food.

- Example: Tigers, Lions, etc.

Decomposers

Decomposers attack the dead bodies of producers and consumer and decompose them into simpler compounds. During the decomposition inorganic nutrients are released. These inorganic nutrients together with other organic substances are then utilized by the producers for the synthesis of their own food.

- Example: Microorganisms like Bacteria and Fungi.

Bacteria and fungi are decomposers. They eat decaying matter dead plants and animals and in the process they break them down and decompose them. When that happens, they release nutrients and mineral salts back into the soil which then will be used by plants.

1.4 Oxygen Cycle and Nitrogen Cycle

Oxygen Cycle

The oxygen cycle is the bio-geo-chemical cycle that describes the movement of oxygen within the three main reservoirs:

- The atmosphere (air).
- The global sum of all ecosystems.
- The lithosphere.

Failures in the oxygen cycle is within the hydrosphere can result in the development of hypoxic zones. The main driving factor of the oxygen cycle is photosynthesis, which produces sugars and free oxygen from carbon dioxide and water.

$$6CO_2 + 6H_2O + \text{energy} \rightarrow C_6H_{12}O_6 + 6O_2$$

An additional source of atmospheric free oxygen comes from photolysis, whereby high-energy ultraviolet radiation breaks down atmospheric water and nitrous oxide into component atoms. The free H and N atoms escape into space, leaving O_2 in the atmosphere.

$$2H_2O + \text{energy} \rightarrow 4H + O_2$$
$$2N_2O + \text{energy} \rightarrow 4N + O_2$$

It also consumes free oxygen by chemical weathering and surface reactions. An example of surface weathering chemistry is formation of iron oxides (rust).

$$4FeO + O_2 \rightarrow 2Fe_2O_3$$

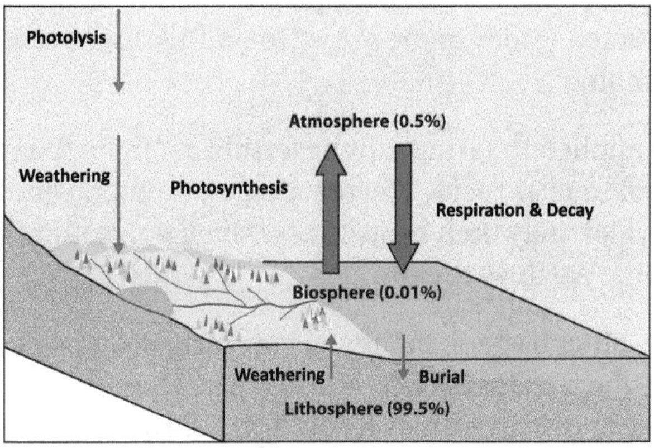

Oxygen cycle.

Nitrogen Cycle

The main component of nitrogen cycle starts with the element nitrogen in the air. Two nitrogen oxides are found in air as a result of interactions with the oxygen. Nitrogen will only react with oxygen in the presence of high temperatures and pressures found near the lightning bolts and in combustion reactions in power plants or internal combustion engines.

Nitric oxide, NO and nitrogen dioxide, NO_2, are formed under these conditions. Eventually nitrogen dioxide may react with water in the rain to form Nitric acid, HNO_3. Nitrates thus formed may be utilized by plants as a nutrient.

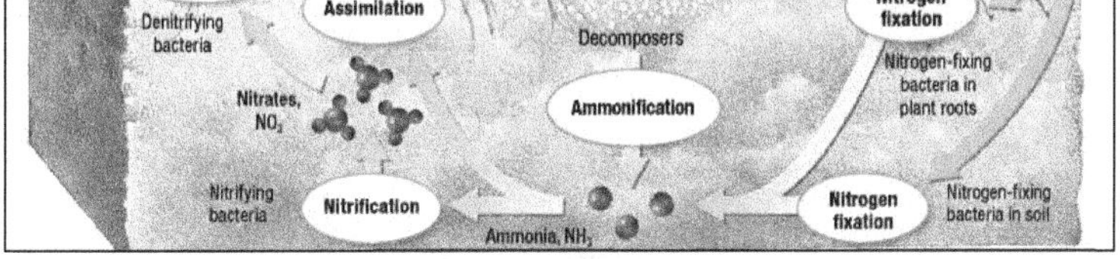

Nitrogen cycle.

Nitrogen in the air becomes a part of biological matter mostly through the actions of bacteria and algae in a process called as nitrogen fixation. Legume plants such as alfalfa, clover and soybeans form nodules on the roots where nitrogen fixing bacteria take nitrogen from the air and convert it into ammonia, NH_3.

The ammonia is further converted by other bacteria first into nitrite ions, NO_2^- and then into nitrate ions, NO_3^-. Plants utilize the nitrate ions as the nutrient or a fertilizer for growth. Nitrogen is incorporate in many amino acids which are further reacted to make proteins.

Ammonia is also made through a synthetic process called the Haber Process. Nitrogen and hydrogen are reacted under great pressure and temperature in the presence of a catalyst to make ammonia.

Ammonia is directly applied to farm fields as fertilizer. On further processing with oxygen, it could be turned to nitric acid. The reaction of ammonia and nitric acid produces ammonium nitrate which may then be used as a fertilizer. Animal wastes when decomposed also return to the earth as the nitrates.

To complete the cycle, other bacteria in the soil may carry out a process known as denitrification which converts the nitrates back to nitrogen gas. The side product of this reaction is the production of a gas known as nitrous oxide, N_2O. Nitrous oxide, also known as "laughing gas" - mild anesthetic, is also a greenhouse gas which contributes to global warming.

1.5 Energy Flow in the Ecosystem

Solar energy is transformed to chemical energy in photosynthesis by plants. Some of the chemical energy used by the plants for their growth and remaining is transferred to other tropic levels.

Tropic levels are feeding levels or various steps through which the energy passes in an ecosystem. The plants are used by the helium and heliotaxes are used by carnivores as their food. Finally plant, herbivores are decomposed and then the energy goes to the environment.

Sun → Plants → Animals → Bacteria

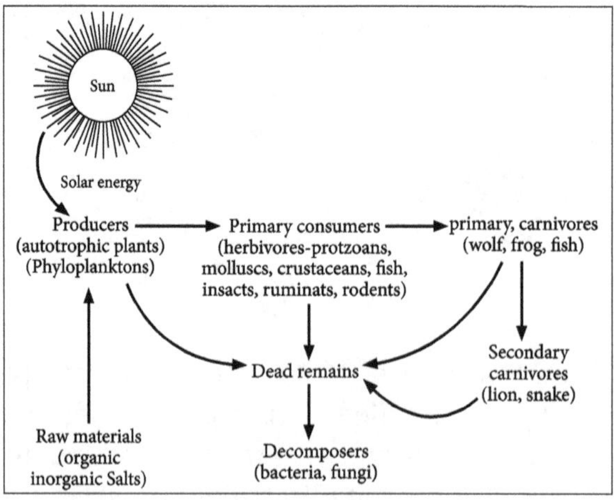

Energy flow at different levels of Ecosystem.

1.5.1 Ecological Succession Processes

The progressive replacement of one community by another life along the development of stable community in a particular area is called ecological succession.

Stages of ecological succession:

- Pioneer community: The first group of organism, which establishes their community in the area, is called "Pioneer Community."
- Seres or Seral Stage: The various developmental stages of the community are called as "seres".
- Community: It is the group of plants or animals living in an area.

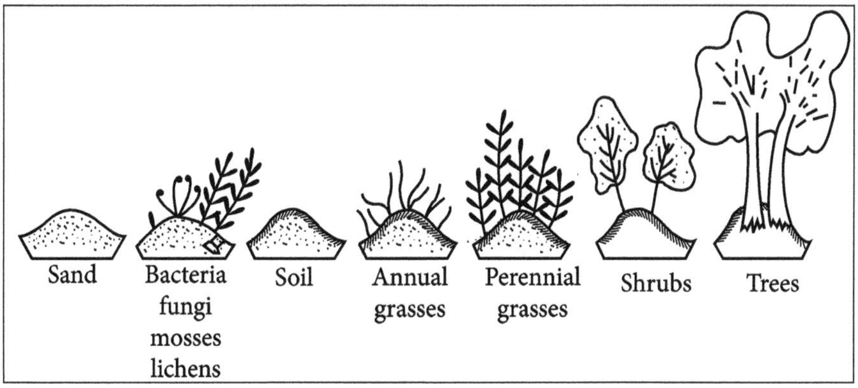

Ecological succession.

Types of Ecological Succession

Ecologists recognize two types of ecological succession, based on the conditions present at the beginning of the process such as:

Primary Succession: It involves the gradual establishment of biotic communities on the lifeless ground.

- Xerarch**:** Establishment starts in the dry area like desert and rock.
- Hydrarch (or) Hydrosere: Establishment starts in the watery area like pond and lake.

Secondary Succession: It involves the establishment of biotic communities in an area, where some type of biotic community is already present.

Process of Ecological Succession

The process of ecological succession can be explained in following steps:

- Nudation: It is the development of a bare area without any life form.

- Invasion: A biological invader is a species of animal, plant or microorganism which most usually transported in or intentionally by man and spreads into new territories. Some distance from its home territory. If one looks at long time scales then invasions are probably a frequent phenomenon. However, human activity has increased the frequency of invasions dramatically by disrupting biogeographic barriers and increasing exchanges.
  - Migration: Migration of seeds is brought about by water, wind or birds.
  - Establishment: The seed then germinate and grow on the land and establishes their pioneer communities.

Competition: As the number of individual species grows, there arises competition with the same species and between different species for water, space and nutrients.

Reaction: The living organism which takes water, nutrients and grows and then modifies the environment is known as reaction. This modification becomes unsuitable for the existing species and also favors some new species, which replace the existing species.

Stabilization: It leads to stable community, which is in equilibrium with the environment.

1.5.2 Forest Ecosystem

The entire assemblage of organisms together with their environmental substrate, interacting inside a defined boundary. Forests and woodlands occupy about 38% of the Earth's surface and they are more productive and have greater biodiversity than other types of terrestrial vegetation.

The Ecological Benefits

- Production of oxygen: During photosynthesis trees produce oxygen which is essential for life on earth.
- Reducing global warming:
  - The main greenhouse gas carbon dioxide (CO_2) is absorbed by the trees (forests).
  - The trees absorb the main greenhouse gas carbon dioxide (CO_2), which is a raw material for photosynthesis. Thus the problem of global warming causes by greenhouse gas CO_2 is reduced.
- Soil conservation: Roots of trees bind the soil tightly and prevent soil erosion. They also act as wind breaks.

- Regulation of hydro logical cycle: Watersheds in forest act like giant sponges which absorb rainfall, slow down the runoff and slowly release the water for recharge of springs.
- Pollution modulators: Forests can absorb many toxic gases and noises and helps in preventing air and noise pollution.
- Wildlife Habitat: Forests are the homes of millions of wild animals and plants.

Characteristics of Forest Ecosystems

Forests are characterized by warm temperature and adequate rainfall, which make the generation of number of ponds, lakes etc.

- The forest support many wild animals and protect the biodiversity.
- The forest maintains the climate and rainfall.
- The soil is rich in organic matter and nutrients, which support the growth of trees.
- Since penetration of light is so poor, the conversion of organic matter into nutrients is very fast.

Types of Forest Ecosystem

Depending upon the climate conditions, forest can be classified into the following types:

- Tropical deciduous forests.
- Tropical rain forests.
- Temperate rain forests.
- Tropical scrub forests.
- Temperate deciduous forests.

Features of Different Types of Forests

- Tropical Rain Forests: They are found near the equator and are characterized by high temperature. They have broad leaf trees like teak and sandal and the animals like lion, tiger and monkey.
- Tropical Deciduous Forests: They are found little away from the equator. They are characterized by warm climate and rain is only during the monsoon. They have different types of deciduous trees like maple, oak and animals like deer, fox, rabbit and rat.

- Tropical Scrub Forests: These are characterized by a dry, climate for longer time. They have small deciduous trees and shrubs and animals like deer, fox, etc.
- Temperate Rain Forests: They are found in temperate areas with adequate rainfall. They are characterized by coniferous trees like pines, red wood, firs, etc. and animals like squirrels, fox, cats, bear, etc.
- Temperate Deciduous Forests: They are found in areas with moderate temperatures. They have major trees including broad leaf deciduous trees like oak, hickory and animals like deer, fox, bear, etc.

Structure and Function of Forest Ecosystem

Abiotic Components

The abiotic components are inorganic and organic substances found in the soil and atmosphere. In addition to minerals, the occurrence of litter is characteristic features of majority of forests.

- Examples: Climatic factors and minerals.

Biotic Components

Producers

The plants absorb sunlight and produce food through photosynthesis.

- Examples: Trees, Shrubs and ground vegetation.

Consumers

- Primary Consumers: They directly depend on the plants for their food.

  Examples: Ants, flies, insects, mice, deer, squirrels.

- Secondary Consumers: They directly depend on the herbivores for their food.

  Examples: Snakes, birds, fox.

- Tertiary Consumers: They depend on the primary carnivores for their food.

  Examples: Animals like tiger, lion, etc.

Decomposers

They decompose the dead plant and animal matter. Rate of decomposition in tropical and subtropical forests is more rapid than in the temperate forests. Examples: Bacteria and fungi.

1.5.3 Grassland Ecosystem

Grasslands occupy about 19% of the earth's surface. The major grassland ecosystems of the world are the great plains of Canada and United States, Argentina to Brazil and Asia to Central Asia.

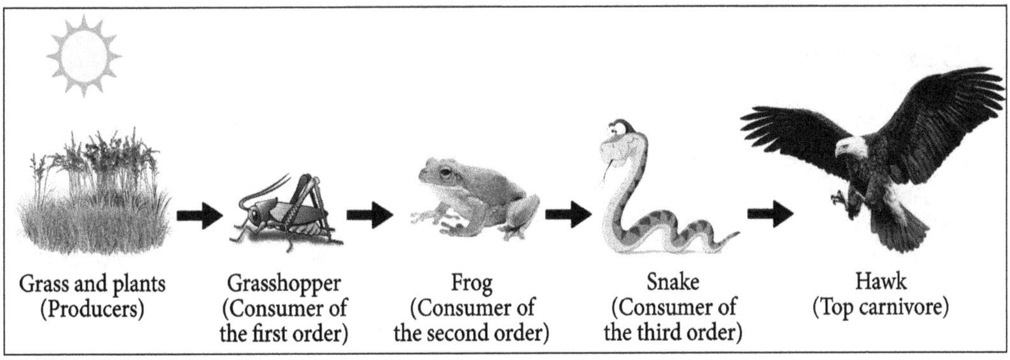

Food chain in grassland ecosystem.

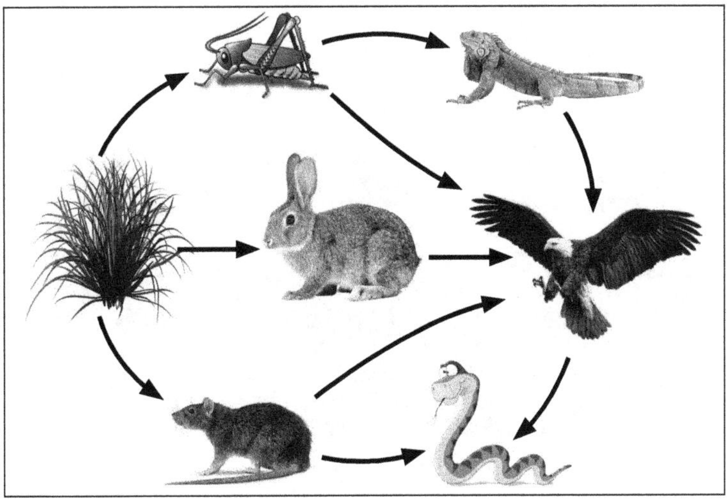

Food web in a grassland ecosystem.

The various components of a grassland ecosystem are as follows:

- Abiotic substances: These include the nutrients present in the soil and the aerial environment. The elements required by plants are hydrogen, oxygen, nitrogen, phosphorous and sulphur. These are supplied by the soil and air in the form of water, CO_2, phosphates, nitrates and sulphates. In addition to these some trace elements are also present in the soil.
- Primary producers: These are mainly grasses of the family, Graminae, large variety of herbs, some shrubs and scattered trees.
- Consumers: Herbivores such as grazing mammals, insects, some termites and millipedes are primary consumers. The animals like fox, jackals, snakes, frogs,

lizard, birds etc., are the carnivores feeding on herbivores. These are the secondary consumers of the grassland ecosystem. Hawks occupy the tertiary trophic level as these feed on the secondary consumers.

- Decomposers: These include bacteria, molds and fungi. These bring the minerals back to the soil to be available to the producers again.

1.5.4 Desert Ecosystem

One can find at least one desert on every continent except Europe and Antarctica. In order to be considered as a desert, it must receive less than 10 inches of water per year.

There are plenty of differences between the deserts of world. Some deserts are made of fine, red sand; others consist of sand mixed with pebbles and rocks. The desert sand started out as rock but after some years, weathering by wind and water has created dunes in the deserts. These sands are mostly minerals and sometimes oil can be found hidden deep within the rocks.

Desert Ecosystem.

Structure and Function

The various components of a desert ecosystem are:

- Abiotic Component: The abiotic component includes the nutrients present in the soil and in aerial environment. The characteristic feature of the abiotic component is the lack of organic matter in the soil and scarcity of water.

- Biotic Component: The various biotic components representing two functional groups are:

Producer Organisms:

- Producers are mainly shrubs or bushes, some grasses and a few trees. Surprisingly, there are many species of plants that can survive in the desert. Most of them are succulents, which mean they store water. Others have the seeds that lay dormant until a rain awakens them. Regardless, these plants find a way to get water and protect them from the heat.

- The most famous desert plant is the cactus. There are many species of cacti. The saguaro cactus is the tall, pole shaped cactus. The saguaro can grow to a height of 40 feet. It can hold several tons of water inside its soft tissue. Like all other cacti, saguaro has a thick, waxy layer that protects it from the Sun.
- Other succulents include the desert rose and the living rock. This strange plant looks like a spiny rock. It's disguise protects it from predators. The welwitschia is a type of weird looking plant. It has two long leaves and a big root. This plant is actually a type of tree and it can live for thousands of years.
- There are many other kinds of desert plants. Some of them have beautiful flowers; others have thorns and deadly poisons. Even in worst conditions, these plants continue to thrive.

Consumers: These include animals such as insects and reptiles. Besides them, some rodents, birds and some mammalian vertebrates are also found.

1.5.5 Aquatic Ecosystems

The aquatic ecosystem deals with water bodies. The major types of organisms found in aquatic environments are determined by the water's salinity:

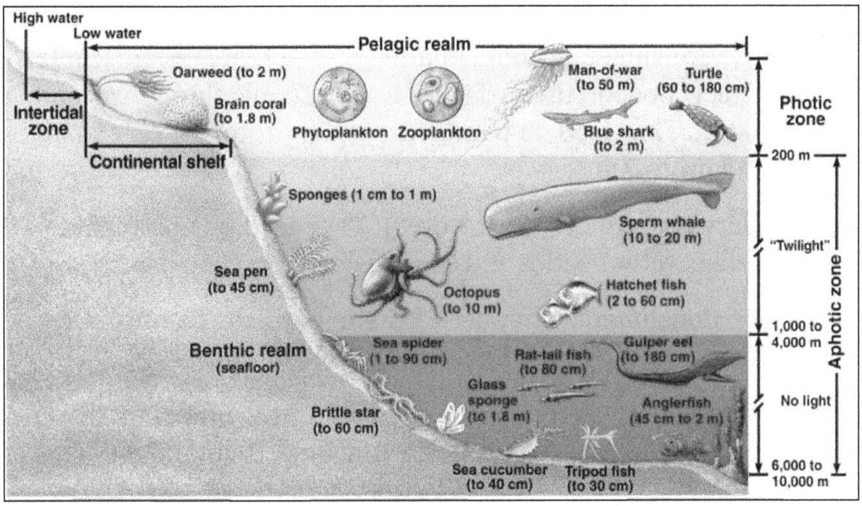

Aquatic Ecosystem.

Types of Aquatic Life Zone

Aquatic life zone is divided into two types:

- Fresh water life zone. Example: Ponds, streams, lakes, rivers.
- Salt water life zone. Example: Oceans, estuarine.

Pond Ecosystem: A pond is a fresh water aquatic ecosystem where water is stagnant. It receives enough water during rainy season. It contains several types of algae, aquatic plants, insects, fishes and birds.

Characteristics:

- Pond is temporary, only seasonal.
- It is stagnant fresh water body.
- Ponds get polluted easily due to limited amount of water.

Structure and Function

Abiotic components:

- Ex: temperature, light, water and organic and inorganic compounds.

Biotic components:

- Producers: They include green photosynthesis organisms. They are of two types:
    - Phytoplankton: There are microscopic aquatic plants, which freely float on the surface of water. Example: Algae and small floating plants like volvox.
    - Microphytes: Example: Large floating plants and submerged plants.
- Consumers:
    - Primary Consumers (Zooplanktons): There are microscopic organisms which freely float on the surface of water. Zooplanktons are found with Phytoplankton. They feed on plants. Example: Protozoa, very small fish, flagellates and protozoans.
    - Secondary Consumers (Carnivores): They feed on zoo-plankton. Example: Insects like water beetles and small fish.
    - Tertiary Consumers: They feed on smaller fish. Example: Large fish like game fish.
- Decomposers: They decompose the dead plant and animal matter and their nutrients are released and secured by green plants. Example: Fungi, bacteria and flagellates.

1.6 Biodiversity

Biodiversity is defined as the variety and variability among all groups of living organism and the ecosystem in which they occurs.

Classification of Biodiversity

Biodiversity is generally classified into three types:

- Genetic diversity.
- Species diversity.
- Community or Ecosystem diversity.

1.6.1 Genetic, Species and Ecosystem Diversity

Genetic Diversity

A species with different genetic characteristics are known as subspecies or genera.

Genetic Diversity is Diversity within Species: Within individual species, there are number of varieties which are slightly different from one another. These differences are due to the differences in the combination of gases. Genes are basic units of hereditary information transmitted from one generation to other.

Examples:

- Rice Varieties: All rice varieties belong to species "Organization." But there are, thousands of rice varieties, which show variation at genetic level differ in their size, shape, color and nutrient content.
- Teak Wood Varieties: There are number of teak wood varieties found available. Example: Indian teak, Burma teak, Malaysian teak, etc.

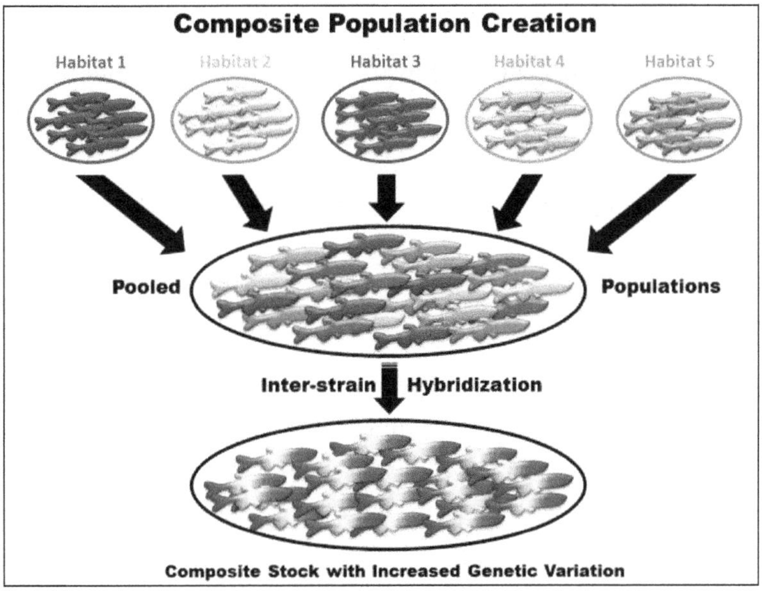

Genetic diversity.

Species Diversity

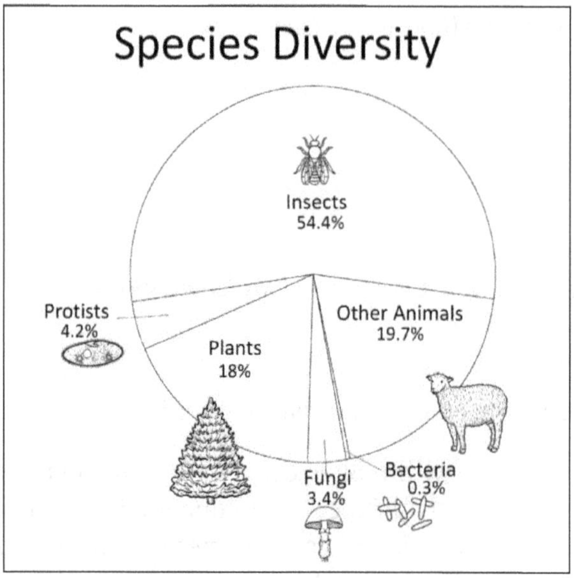

Species diversity Graph.

A discrete group of organisms of the same kind is known as species. Species diversity is the diversity between differed species. The sum of varieties of the entire living organism at the species level is known as species diversity.

Biotic component is composed of a large number of species of plants, animals and microorganisms which interact with each other and with the abiotic component of the environment.

Example:

- Total numbers of living species in the earth are more than 20 million. But, of which only about 1.5 million living organisms are found and given scientific names.
- Plant species: Apple, mango, grapes, wheat, rice, etc.
- Animal species: Lion, tiger, elephant, dear, etc.

Community Ecosystem Diversity

It is a set of biotic components interacting with one another and with abiotic components. The diversity at the ecological or habitat level is known as ecosystem diversity. A large region with various ecosystems can be considered as ecosystem diversity.

Example:

- River ecosystem: The River which include the fish, aquatic insects, muscles and variety of plants that have adapted. Thus, the ecosystem diversity is a aggregate of different environment types in a region. It explains the interaction between living organisms and physical environment in ecosystems.

1.7 Bio-Geographical Classification of India

Bio-geographic classification of India is the division of India according to bio-geographic characteristics. Bio-geography is the study of the distribution of species (biology) organisms and ecosystems in geographic space and through geological time. There are ten bio-geographic zones in India:

- Trans Himalayan zone.
- Himalayan zone.
- Desert zone.
- Semiarid zone.
- Western ghat zone.
- Deccan plateau zone.
- Gangetic plain zone.
- North east zone.
- Coastal zone.
- Islands present near the shore line.

1.8 Value of Biodiversity

Biosphere is a life supporting system to the human beings. It is the combination of different organisms. Each organism in the biosphere has its own significance. Biodiversity a vital for healthy biosphere. Biodiversity is must for the stability and proper functioning of the biosphere. We get benefits from other organisms in number of ways. Sometimes we realize the real value of the organism only after it is lost in this earth.

The values of biodiversity have been classified into 6 types:

Consumptive use Value

These are direct use values where the biodiversity products are harvested and consumed directly. Example: Food, Drug, Fuel, etc.

Food: A large number of wild plants are consumed by human beings as food. Nearly 80 to 90% of our food crops have been domesticated only from the tropical wild plants. A large number of wild animals are also consumed as food.

Example:

- In Himalayan region.
- Insects: Spider and wild herbivores are consumed by many tribal and nontribal Communities in India.

Drugs: Around 70% of modern medicines are derived from plant and plant extracts. 20,000 plant species are believed to be used medicinally, particularly in the traditional system of Ayurveda and Sidha.

Example:

- Germany alone use more than 2,500 Species of plants for medicinal purposes in Homeopathy and other systems of medicines.
- India uses 3000 species of plants in Ayurveda, Homeopathy and System of medicines.

Fuel: Fire woods are directly consumed by villagers or tribals. The fossil fuels like coal, petroleum and natural gas are also the products of fossilized biodiversity.

Productive use Values

Biodiversity products have obtained commercial values. These products are marketed and sold. These products may be derived from the animals and plants:

Animal Product	Animal
Silk	Silk-worm
Wool	Sheep
Leather	All animals
Food	Fish and animals

Table: Plant Products for Various Industries.

Plant Product	Industry
Wood	Paper and pulp industry, plywood industry
Cotton	Textile industry
Fruits, vegetables	Food industry
Leather	Leather industry

- Rice accounts for 22% of the cropped area and cereals accounts for 39% of the cropped area.

- Oil seed production also helped in saving large amount of foreign exchange spend on importing edible oils.

Social Values

Social value of biodiversity refers to the manner in which the bio resources are used to the society. They are associated with the social life, religion and spiritual aspects of the people.

Example:

- Holy Plants: Many plants are considered as the holy plants in our country. The leaves, fruits of these plants are used in worship. Example: Lotus.
- Holy Animals: Many animals are also considered as holy animals in our country. Example: Cow, Snake, Bull, Peacock.

Ethical Value

- It involves ethical issues like all life must be preserved. In India and in other countries biodiversity is considered to have great value on religions and cultural basis.
- Our rich heritage teaches us to worship plants, animals, rivers and mountains. The ethical value means that a species may or may not be used, but its existence in the nature gives us pleasure.

Aesthetic Value

The beautiful nature of plants and animals insist us to protect the biodiversity. The most important aesthetic value of biodiversity is ecotourism.

Example:

- Eco-tourism: People from far place spend a lot of time and money to visit the beautiful areas, where they can enjoy the aesthetic value of biodiversity. This type of tourism is known as ecotourism.
- The pleasant music of wild birds.
- Color of butterfly, color of flowers, and color of peacocks are very important for their aesthetic value.

Option Values

The option values are the potential of biodiversity that are presently unknown and need to be known. The optimal value of biodiversity suggests that any species may be proved to be a valuable species after some day.

Example:

- The growing biotechnology field is searching a species for causing the disease of cancer and AID's.
- Medicinal plants and herbs play a very important role in our Indian economic growth.

1.8.1 Biodiversity at Global, National and Local Levels

Biodiversity Levels

Biosphere is a life supporting system to the human beings. It is the combination of different organisms. Each organism in the biosphere has its own significance. Biodiversity it's vital for healthy biosphere. Biodiversity is must for the stability and proper functioning of the biosphere.

Global Biodiversity

Total numbers of living species in the world are about 20 million; about 1.5 million species are found and given scientific names. Tropical deforest 0.5% every year.

Terrestrial Biodiversity or Biomass

It is the largest ecological units present in different geographic areas.

Tropical Rain Forests

Earth's largest storehouse of biodiversity, they are inhabited by millions of species of plants, insects, birds, amphibians and mammals 50 to 75% rain forests.

- Medicinal Plants: More than 25% of the world's prescription drugs are extracted rain forest.
- Flowering Plant: It has been estimated that nearly 1, 30, 000 flowering plant species are 1-3% plant species.

Protection of Tropical Rain Forests: It is essential to project our tropical rain forests. The silent valley in Kerala is only place in India where tropical rain forests available.

Temperature Forests

These have much less biodiversity globally. They have nearly:

- 1, 70, 000 flowering plants.

- 30, 000 vertebrates.
- 2, 50, 000 other groups of species.

Marine Diversity

Marine diversity is much higher than terrestrial biodiversity, but is less known described. Estuaries coastal waters and oceans are biologically diverse but the diversity is very low sea is the cradle of every known phylum. 35 existing of cellular animals 34 are marine.

1.8.2 India as a Mega-Diversity Nation

Among the biologically rich nations, India stands among the top 10 or 15 countries for its great variety of plants and animals, many of which are not found elsewhere. India has 350 different mammals, 1,200 species of birds and 453 species of reptiles and 45,000 plant species, of which most are angiosperms.

These include generally high species diversity of ferns and orchids. India has 50,000 known species of insects, which includes 13,000 butterflies and moths. It is estimated that the number of unknown species could be several times higher. It is estimated that 18% of Indian plants are endemic to the country and found nowhere else in the world.

As one of the oldest and largest agriculture societies, India has also a striking variety of at least 166 species of crop plants and around 320 species of wild relatives of cultivated crops. There is a vital, but often neglected factor when we focus on biodiversity. It may be a matter of surprise for many to understand that the who officially constitute 7.5 % of India's population have preserved about 90 % of the country's diversity. To a large extent, survival of our biodiversity depends on how best they are looked after.

1.8.3 Hot-Spots of Biodiversity

The hot spots are the geographic areas which possess the high endemic species. At the global level, these are the areas of high conservation priority. If these species lost, they can neither be replaced nor regenerated.

Criteria for Recognizing Hot Spots

- The hot spots should have a significant percentage of specialized species.
- The richness of the endemic species is the primary criterion for recognizing hot spots.
- The site is under threat.

- It should contain important gene pools of plants of potentially useful plants.

Reasons for Rich Biodiversity in the Tropics

The following are the reasons for the rich biodiversity in the tropics:

- Warm temperatures and high humidity in the tropical areas provide favorable conditions.
- The tropics have a more stable climate.
- Among plants, rate of outcrossing appear to be higher in tropics.
- No single species can dominate and thus there is an opportunity for many species to coexist.

Hot spots of Biodiversity in India

- Western Ghats - Sri Lanka Region.
- Eastern Himalayas - Indo-Burma Region.

Eastern Himalayas: Geographically this area comprises of Nepal, Bhutan and neighboring states of Northern India. There are 35,000 plant species found in the Himalayas, of which 30% are endemic. Eastern Himalayas are also rich in wild plants of economic values. Examples: Rice, Banana, Juice and Sugarcane.

The taxol yielding plant is also sparsely distributed in the region:

- 63% mammals are from Eastern Himalayas.
- 60% of the Indian Birds are from North West.
- Huge wealth of fungi, insects, mammals, birds have been found in this region.

Western Ghats: The area comprises Maharashtra and Kerala. Nearly 1500 endemic, dicotyledons plant species are found from Western ghats. 62% amphibians and 50% lizards are endemic in Western Ghats. It is reported that only 6.8% of the original forests exist today while the rest are deforested or degraded.

- Some common plants: Japonica, Rhododendron and Hypericum.
- Some common animals: Blue bird, Lizard, hawk.

1.8.4 Threats to Biodiversity: Habitat Loss, Poaching of Wildlife and Man-Wildlife Conflicts

Extinction is the Complete Elimination of Wild Species

It is a Natural but slow process but due to unplanned activities of man; the rate of decline of wild life has been particularly rapid in the last one hundred years. There are a number of causes which are known to cause the extinction of wildlife.

Wetlands are drained to increase the agricultural land. These changes have economic implications in long term. The current destruction of the remaining large areas of wilderness habitats, especially the super diverse tropical forests and coral reefs, is the most important threat worldwide to biodiversity. Scientists have estimated that the human activities are likely to eliminate almost 10 million species by the year 2050. There are around 1.8 million species of plants and animals, large and microscopic, known to science in the world at present.

The number of species however is likely to be greater by a factor of at least 10. The plants and insects as well as other forms of life are not known to science are continually being identified in the worlds' 'hotspots' of diversity. Unfortunately at the present rate of extinction, about 25% of the worlds' species will undergo extinction fairly rapidly. This may occur at the rate of 10 to 20 thousand species per year, thousand to ten thousand times faster than the expected natural rate.

Much of this mega extinction spasm is related to human population growth, industrialization and changes in land use patterns. A major part of these extinctions will occur in 'bio-rich' areas such as tropical forests, wetlands and coral reefs. Loss of wild habitats due to rapid human population growth and the short term economic development are one of the major contributors to the rapid global destruction of biodiversity.

Habitat Loss

It is the most serious threat to wildlife. It is due to Environmental pollution. Habitat loss also results from man's introduction of species from one area into another, disturbing the balance in existing communities. Loss of species occurs due to destruction of natural ecosystems, either by over extraction of their resources or through pollution of air, water and soil or for conversion to agriculture or industry.

Effects of Habitat Loss on Biodiversity

Habitat loss is a process of environmental change in which a natural habitat is rendered functionally unable to support the species present. It may be natural or unnatural and may be caused by habitat fragmentation, geological processes, climate change or human activities such as the introduction of invasive species or ecosystem nutrient depletion. In the process of habitat destruction, the organisms that previously used the site are displaced or destroyed, reducing biodiversity.

Human destruction of habitats has accelerated greatly in the later-half of twentieth century. Natural habitats are often destroyed through human activity for the purpose of harvesting natural resources for industry production and urbanization.

Clearing habitats for agriculture, for example, is the principal cause of habitat destruction. Other important cause of habitat destruction includes mining, logging and urban sprawl. Habitat destruction is currently ranked as the primary cause of species extinction worldwide.

Consider the exceptional biodiversity of Sumatra. It is home to one sub-species of orangutan, a species of critically endangered elephant and the Sumatran tiger. However half of Sumatra's forest is now gone. The neighboring island of Borneo which is the home to the other sub species of orangutan has lost a similar area of forest and forest loss continues in the protected areas.

The orangutan in Borneo is listed as endangered by the International Union for Conservation of Nature (IUCN), but it is simply the most visible thousands of species which does not survive on the disappearance of the forests of Borneo. The forests are being removed for their timber and to clear the space for plantations of palm oil, oil used in Europe for many items including food products, cosmetics and bio-diesel.

Man Wild-Life Conflicts

Wildlife protecting bodies and local people has undergone indirect conflicts. Local communities have been affected due to: Loss of wildlife resources and Depletion of income.

Depletion of Income

Income has depleted of rural population due to check on hunting, harvesting and human conflict is an increasing example. Elephants being cute and lovable are the cherish able animals. But if they are put next to our house they turbo violent leading to conflict with human.

Biodiversity of an area influences every aspect of lives of the people who inhabit it. Their livelihood and their living space depend on the type of ecosystem. Even people living in urban areas are dependent on the ecological services provided by the wilderness in the PAs. It is linked with the service that nature provides us. The quality of water we drink and use, the air we breathe, soil on which our food grows are all influenced by a wide variety of living organisms both plants and animals and the ecosystem of which each species is linked with in nature.

While it is well known that plant life removes the carbon dioxide and releases oxygen we breathe, it is less obvious that fungi, small soil invertebrates and even microbes are essential for plants to grow. A natural forest maintains the water in the river after the monsoon or the absence of ants could destroy life on earth, are to be appreciated to understand how we are completely dependent on the living 'web of life' on earth. This includes mankind as well. Think about this and we cannot but want to protect out earth's unique biodiversity. We are highly dependent on these living resources.

1.8.5 Endangered and Endemic Species of India

According to International Union of Conservation of Nature and Natural Resources (IUCN), the species are classified into the following types:

- Extinct Species: A species is declared as extinct, when it is no longer found in the world.

Endangered species.

- Endangered Species: A species is declared as endangered, when its number has been reduced to a critical level. Unless it is under protection and conserved, it is in immediate danger of extinction.
- Vulnerable Species: A species is said to be vulnerable when its population is facing continuous decline due to habitat destruction of over exploitation. Such a species is still abundant.
- Rare Species: A species is said to be rare, when it is localized within restricted area (or) they are thinly scattered over a more extensive area. Such species are not endangered or vulnerable.

1.8.6 Conservation of Biodiversity: In-Situ and Ex-Situ Conservation of Biodiversity

Conservation is the protection, preservation, management or restoration of wildlife and natural resources such as forests and water. Through the conservation of biodiversity, the survival of many species and habitats which are threatened due to human activities are ensured. Other reasons for conserving biodiversity includes securing valuable natural resources for future generations and protecting the well-being of ecosystem functions.

In-Conservation

In-conservation involves the protection of flora and fauna within its natural habitat, where the species normally occurs is called in-conservation. The natural habitats or ecosystems maintained under in-conservation are called "protected areas".

Important in-Conservation

- Biosphere reserves,
- National parks,
- Wildlife sanctuaries,
- Gene sanctuary, etc.

Methods of in-Conservation

Around 4% of the total geographical area of the country is used for in-conservation. The following methods are presently used for in-conservation.

In-conservation	No. available
Biosphere reserves	7
National parks	80
Wild-life sanctuaries	420

Biosphere Reserves

Biosphere reverses cover large area, more than 5000 sq. km. It is used to protect species for long time. Some important Biospheres Reserves in India.

Table: Biospheres Reserves in India.

S.No.	Name	Area of Biosphere (sq km)	Date of establishment	District	State
1	Agasthyamalai	1701.00	2001	Kollam, Thiruvananthapuram	Kerala
2	Achanakmar-Amarkantak	3835.51	2005	Anuppur, Dindori & Bilaspur	Madhya Pradesh & Chhattisgarh
3	Dibru-Saikhowa	765.00	1997	Dibrugarh and Tinsukia	Assam
4	Dehang-Debang	5111.5	1998	Siang & Debang Valley	Arunachal Pradesh

5	Gulf of Mannar	10,500.00	1989	Indian part of Gulf of Mannar	Tamil Nadu
6	Great Nicobar	885.00	1989	Southern-most Island of Andaman and Nicobar	Andaman and Nicobar
7	Manas	2837.00	1989	Part of Kokrajhar, Bongaigaon, Barpeta, Nalbari, Kamrup and Darang	Assam
8	Kanchanjunga	2619.92	2000	Kanchanjunga Hills	Sikkim
9	Nilgiri	5520.4	1986	Part of Wayanad, Bandipur and Nagarhole,Nilambur, Silent Valley and Siruvani Hills	TamilNadu, Kerala and Karnataka
10	Nanda Devi	5860	1988	Chamoli, Almora and Pithoragarh	Uttaranchal
11	Pachmarhi	4926.00	1999	Betul, Hoshangabad and Chhindwara	Madhya Pradesh
12	Nokerek	80.00	1988	Part of Garo Hills	Meghalaya
13	Sunderbans	9630.00	1989	Delta of Ganges and Brahmaputra	West Bengal
14	Similipal	4374.00	1994	Mayurbhanj	Orissa

Role of Biosphere Reserves:

- It gives long-term survival of evolving ecosystem.
- It protects endangered species.
- It serves as site of recreation and tourism.
- It is also useful for educational and research purposes.

Restriction:

No tourism and explosive activities are permitted in the biosphere reserves.

National Park

A National Park is an area dedicated for the conservation of wildlife along with its environment. It is usually a small reserve covering an area of about 100 to 500 sq. ft. Within the biosphere reserves, one or more national parks are also exists.

Role of a National Park:

- It is used to protect, propagate and develop the wildlife.
- It is used for enjoyment through tourism, without affecting the environment.

Restrictions:

- Grazing of domestic animals inside the national park is prohibited.
- All private rights and forestry activities are prohibited within a national park.

Wildlife Sanctuaries

A wildlife sanctuary is an area, which is reserved for the conservation of animals only. At present, there are 492 wildlife sanctuaries in our country.

Role of Wildlife Sanctuaries:

- It allows the operations such as collection of forests products, harvesting of timer, private ownership rights and forestry operations provided.
- It does not affect the animals adversely.
- It protects animals only.

Restrictions

Killing, hunting, shooting or capturing of wildlife is prohibited except under the control of higher authority.

Gene Sanctuary

A gene sanctuary is an area, where the plants are conserved.

Example: In Northern India, two gene sanctuaries are found available.

- One gene sanctuary for citrus.
- One gene sanctuary for pitcher plant.

Advantages of in-Conservation

- It is cheap and convenient method.
- The species gets adjusted to the natural disasters like drought, floods, and forest fires.

Disadvantages

- Maintenance of the habitats is not proper, due to shortage of staff and pollution.

- Large surface area of the earth is required to preserve the biodiversity.

Ex-Conservation

Ex- conservation involves protection of fauna and flora outside the natural habitats.

This type of conservation is mainly done for conservation of crop varieties and the wild relatives of crops.

Role of Ex-Conservation

- It identifies those species which are at more risk of extinction.
- It involves maintenance and breeding of the endangered plant and animal species under controlled conditions.
- It prefers the species, which are more important to man in near future among the endangered species.

Important Ex-Conservation

- Botanical gardens,
- Seed banks,
- Microbial culture collections,
- Tissue and cell cultures,
- Museums.

Methods of Ex-Conservation

The following important gene bank (or) seed bank facilities are used in Ex-conservation:

National Bureau of Plant Genetic Resources (NBPGR):

- NBPGR is located in New Delhi. It uses the preservation techniques to preserve the agricultural and horticultural crops.
- Preservation Technique: It involves the preservation of seeds, pollen of some important agricultural and horticultural crops by using liquid nitrogen at a temperature as low as - 196 °C. Varieties of rice, turnip, radish, onion, have been preserved successfully in liquid nitrogen for several years.

National Bureau of Animal Genetic Resources (NBAGR):

- It preserves the semen of domesticated bovine animals.

National Facility for Plant Tissue Culture Repository (NFPTCR):

- It develops the facility for conservation of varieties of crop plants or tree by tissue culture. This facility has been created within the NBPGR.

Advantages of Ex-Conservation

- Survival of endangered species is increasing due to special care and attention.
- It captive breeding, animals are assured food, water, shelter and also security and hence longer life Span.
- It is carried out in cases of the endangered species, which do not have any chances of survival, in the wild.

Disadvantages of Ex-Conservation

- The animals cannot survive on natural environment.
- It is expensive method.
- The freedom of wildlife is lost.

1.8.7 Field Study of Common Plants, Insects and Birds

These taxa have been selected as they occur nearly everywhere. While one may have to visit a national park or sanctuary to see mammals or reptiles, several plants, Insects and birds can be seen around an urban or rural setting where there is some vegetation. If we have an opportunity to visit a national park or sanctuary, we can add other animals. One needs a little equipment—a Journal to take notes, preferably a pair of binoculars and field guides to identify plants, insects, birch, reptiles and mammals. These are available from Bombay Natural History Society (BNHS).

Plants:

- Identify and list common plant species at the study site.
- Identify if there are rare species by using a field guide a asking a botanist.
- Identify and list the types or plants—trees/shrubs/climbers/grcund cover: Mtn, grasses. Observe their abundance levels.
- Describe five plant species. Document the characteristic features that help in the identification of the selected species:
    - Describe the specific characteristics of the leaves/flowers/fruit/seeds.

- Describe the plant's role in the ecosystem.
- Now is it used and by whom?
- Is It being collected sustainably or over-harvested?
- Is it common or rare? If it is rare, why?
- Is it a keystone species? If so, why?

Animals:

- Identify list type of animal species.
- Observe level of abundance of each species.
- List characteristic features.
- Role of species in ecosystem.
- Feeding and breeding behaviour etc.
- Flow these species can be protected.

This green world is full of a variety of plants, classified as herbs, shrubs, climbers and trees. A tree can be defined as a big, tall, plant, with a woody stem and a rough bark, branching at the top, with leaves at the ends of its twigs. A tree is generally more than three meters tall, but some exceptions also exist, such as palms, which do not branch at all, even though they are trees. In our vicinity we come across different types of trees which are loosely classified as avenue trees, garden trees, forest trees, village trees and so on. The type of tree also varies according to the habitat, climate and soil of the area. Trees of hilly tracts are different from trees growing in the desert.

Sometimes we plant around them in our neighbourhood keeping in mind their uses and importance. In this module we will learn about some common village trees.

Acacia or the Egyptian Thorn Tree

- Acacia is a middle sized, thorny, deciduous tree found in hot low lands. It thrives very well in dry conditions. The bark is almost black, thick, deeply ridged and fissured. The tree has a spreading, open crown of feathery dark green, compound leaves. Its bright yellow flowers appear in the rainy season.
- Its Leaves are twice feathered, 2.5 to 5 cm long dividing into 3 to 6 pairs. Each side stalk bears upto 25 pairs of leaflets. A pair of straight, white spines, about 5 cm in length, is present below each leaf. Leaves are shed in March or April.

- Its Flowers are small bright yellow, clustered tightly to form round, yellow heads. Petals are insignificant. Bright yellow stamens give colour to the heads. The tree flowers from June to August.
- Its Fruits are grayish and are known as pods. Each pod may be upto 20 cm long, deeply pinched between every seed. Each pod appears like a bead necklace. A single pod contains about 8-14 seeds. The fruit is indecent.

Names

Acacia is commonly known as 'Babool' in many parts of India. It is also called 'keekar'. Its English name is Egyptian Thorn Tree. Botanically the plant is known as Acacia nilotica or Acacia arabica. It is a member of the family Mimosaceae.

Uses

Acacia is one of the most important trees of India. It is planted on the boundaries of fields. It provides protection to fields and fodder to goats and camels. The bark and fruits are used for tanning. Gum from this tree is used in calico printing and dyeing. It is also used in making inks, paints and confectionery. The heart wood is hard and is widely used in making agricultural tools. The wood is also used as fuel. In all, the tree is very useful and can be said to be a true friend of farmers.

Botanical Importance

Acacia is indigenous to Sindh, Gujarat and North Deccan and is cultivated everywhere in India, considering its various uses. The tree provides good fodder to herbivores like goats and camels. The flowers are visited by bees for nectar and to pollinate them.

Due to increased urban developmental pressure, this tree is becoming less visible around cities, even though it is resistant to air pollution. The weaver bird or 'baya's' nests are often found dangling from the tree. It is one of the preferred nesting sites of the weaver bird. It also provides food to the larvae of Gram blue and Acacia blue butterflies.

Date Palm

The date palm is a tall, feather-leafed palm. Its stem is slender and curving, with plumore crown. The tree is evergreen. The Trunk is grey-brown, covered by stumps of fallen leaves. It is un-branched, with a crown of leaves at the top. Male and female Plants are separate. This condition is known as 'dioecious'.

Leaves are very long, arching, grayish-green, compound. The leaflets are attached to the rachis. Each leaflet can be upto 45 cm long. Sharp spines are found on the lower portion of the leaf stem.

The Flowers are trimerous i.e. three-petaled. They are fragrant. Male and female flowers grow on different trees. The stalk of flower cluster is flat and thick, like a camel's tail. The flower buds are enclosed in large sheaths. Male flowers are larger, more closely spaced and creamy yellow.

The fruit comes in large clusters. The fruits are green when young but become orange-yellow on maturing and finally turn to a deep red-brown. They are edible, but contain very little quantity of sweet pulp. The seed is large and grooved. The seed is called 'gudhali'.

Uses

The tree is of great economic value, especially for villagers. Leaves are sugary. The freshly tapped juice is rich in sugars and ferments quickly. It is known as 'taadi'. The unfermented fresh juice is very refreshing and sold as 'neera'. The woody trunk is used as beams and support in making houses. The leaf-stalks are used to make paper and rope. Old fallen leaf stalks provide fuel to the poor.

Ecological Importance

The sweet fruits are eaten by many birds. This tree plays a very important role in rural economy. Its multipurpose uses can help in supporting and strengthening sustainable development. The tall trees are a favorite site for nesting weaver birds. One can see many nests of the weaver bird hanging from its leaves. Thus, this tree helps different herbivores, as well as human beings, in many ways. In short it provides all the basic needs of a man i.e. food, clothing and shelter.

Pongam Tree

- Pongam is a familiar medium sized, deciduous tree. It reaches usually upto 10-20 meter high with a spreading, shady crown. It is very common along river banks, nalas and streams. The tree is especially found in Sudarbans. The bark is smooth and gray.
- The leaves are compound. Each leaf consists of about 5-9 leaflets, placed opposite each other. Leaflets are green, oval in shape and 4-10 cm long, with elongated tip. The pale green young leaves make the tree look very beautiful. The mature leaves are attacked by leaf miner insects, making the leaf very ugly. Leaves fall in late winter.
- New leaves appear in April, with a second flush during rains. The leaf miner worm attacks the old leaves in December.
- Flowers come in short clusters. They are pea-like. The flowers are fragrant white of pale pink. The flower has five petals, one of which is hood-like and called the

'standard petal'. The stamens are clustered together. The flowers appear on the tree from April-June, when the tree is nearly leafless.

- Fruits are woody, pod-like, about 5 cm long, almost oval and flattened, with a characteristic pointed beak at the apex. Generally each pod contains one seed. The pods are initially green and become grey-brown on maturity. They are indehiscent and ripen between March to May of the next year.

Names

Pongam is variously named as 'Indian Beech Tree', 'Ponga-oil Tree'. In Hindi, Marathi and Bengali it is known as 'Karanj'. This name is probably associated with the abundance of the tree on the island of Karanjia, near Mumbai. The botanical name is 'Pongamia pinnata'. 'Pongamia' has been coined from the Tamil name of Pongam. This tree belongs to the family Papilionaceae.

Uses

The Pongam is mainly used as a shade tree or as an ornamental plant. It is very suitable for road-side avenues and on railway station platforms, but is more commonly found in villages. Its leaves provide green manure and are used to cure skin diseases.

They are also very effective against white ants. The oil obtained from seeds is used to treat scabies, herpes and rheumatism. Its oil is used as a lubricant, lamp oil, pesticide and in soap making. It is locally used as fuel wood.

Ecological Importance

The leaves are not eaten by goats, so it is planted in large numbers without any protection on roadsides, creating green belts. Though the leaves are not palatable to cattle, the larvae of Chestnut Streaked Sailor Butterfly, Indian Sunbeam Butterfly feed on them freely. The flowers of Pongam also provide food to the larvae of Common Cerulean Butterfly. In this way the tree provides food to various herbivores, including leaf miners.

The Temple Tree

- The temple tree is a small deciduous, low branching tree, generally planted in temples and gardens. The leaves are large, crowded at the end of the branches. The bark is very thin, shiny, greenish-brown and exudes a milky sap when cut or bruised. It is native to a hot, tropical territory, stretching from South Mexico, Panama and West Indies to the North of South America. At present, it grows extensively throughout India.

- It was probably introduced to India by the Portuguese. It was found growing in the Malabar Hills in Mumbai in 1787.

- The leaves are large, broadly lanceolate, tapering at both ends and quite distinctive due to prominent parallel veins. They are about 55 cm long, hairless and smooth when mature. The leaves are darker on the top and dull, without any shine. They are borne in crowded spirals at the end of the branches. Branches are cylindrical in shape.
- The tree remains leafless from December until the onset of the rainy season.
- The flowers are strongly fragrant. They appear in upright clusters at the top of the branches, from a common stalk. The flowers are waxy, white and range from pink to deep crimson. The centre of the flower is generally tinged with golden yellow. Petals are five and twisted, that is, they overlap each other. The flowering period is between March and April and again between July and October. The tree is liked by all because of its sweet-scented flowers, which are available almost throughout the year.
- The fruit is a leathery pod. Pods come in pairs of 12-15 cm. They are deep green, filled with seeds, with a tassel of silk. The tree rarely fruits in our country.

Names

The Temple Tree is known in English as Frangipani. It is also known as the 'Pagoda tree' as it is generally found near pagodas. In Hindi it is called 'Champa'. The botanical name is 'Plumeria acutifolia', named after the French botanist Charles Plumier, while the term 'acutifolia' refers to the pointed, tapering leaves. It is also called 'Plumeria rubra', where 'rubra' refers to the variety with red flowers.

Uses

It is a very attractive tree, loved by gardeners and nature lovers. The sweet scented flowers are offered in worship. The bark relieves fever, heals sores and acts as a powerful purgative. The crushed leaves are applied to the skin to reduce swelling. The milky latex is used along with coconut oil to treat skin troubles.

Ecological Importance

The flowers are visited by a variety of insects, especially bees, which pollinate them. The leaves of trees growing along roadsides are often found covered with heavy dust deposits. In this way the tree helps control particulate pollution in cities. It also seems to be comparatively resistant to air pollution.

The Indian Gooseberry Tree

- The Indian Gooseberry Tree is a medium sized, deciduous tree, generally found in dry deciduous forests, but cultivated for its useful sour fruits on a large scale.

It can also be seen in gardens or in villages, near temples, where it is planted for its religious significance.

- The Indian Gooseberry is a distinctive tree with a thin grey bark which flakes off to reveal a yellow-brown fresh bark beneath. The leaves are very small and generally mistaken as leaflets of a compound leaf. Each leaf is 8-12 mm long, narrow, with pointed apex. The tree remains leafless during January and February. New leaves appear in March. The tree flowers in March as well. Separate male and female flowers are to be found on the Indian Gooseberry. Male flowers are many and found in clusters at the base of the twig, while the female flowers are found towards the apex of the branches. The flowers are small and inconspicuous. Female flowers are pollinated by insects, especially bees. The Indian Gooseberry is known for its fruits, which are round and 2-4 cm in diameter.
- They are yellow-green in color, with a thin, translucent skin. About 6-8 faint vertical lines can be observed on the fruit. The fruit is extremely sour and is a rich source of Vitamin C. They are 20 times richer in Vitamin C than oranges are.

Names

The Indian Gooseberry is also known as Emblic murobalan. In Hindi it is known as 'Amla', while in Sanskrit it is called 'amlak'. Its botanical name is "Phyllanthus emblica'. It belongs to the family Euphorbiaceae.

Uses

The Indian Gooseberry is a very important medicinal tree. Its fruits are used for making 'Triphala churn' and 'Chyavanparash' and various other ayurvedic medicines. They are also used in making pickles, sherbets, hair oils, etc.

Ecological Importance

The fruits of Indian Gooseberry are eaten by wild animals. The flowers provide nectar to many insects. Research conducted in our country suggests that this tree is moderately resistant to the effects of air pollution due to the rich Vitamin C content, which acts as a powerful antioxidant.

Cluster Fig

- The Cluster Fig is a middle-sized deciduous tree. The tree trunk is crooked and has an open, spreading crown. The bark is relatively smooth, grayish-yellow or rusty brown.
- The bark becomes somewhat scaly with age. A milky sap comes out when the tree is bruised or cut. Leaves are shed in January and the tree is leafless till

March. However, the tree may be evergreen if found near streams or ponds. The Cluster Fig is not a 'strangler' like other figs and is not found as en epiphyte on other fig trees. Aerial roots are also absent.

- A native of Africa, this tree has now reached Asia. It can now be found throughout tropical Asia, from India to Australia. The Cluster Fig is not found in the drier parts of India. It is one of the 'panch pallav' trees of Indian culture, the other four being the peepal, mango, Jamun and the Banyan.
- The leaves are simple, 9-13 cm long, thin & with a few irregular teeth. The teeth disappear as the leaf matures. Leaves are leathery, tapering at both ends. The upper surface is dark and the lower is dull or pale. The leaves are arranged alternately on the twigs. They have three prominent yellow veins. The venation is reticulate.
- Flowers are found enclosed in figs. They are very small. Male and female flowers are separate. The figs grow in large clusters directly from the trunk or from the main branches. This is called 'cauliflorae' and is a characteristic of this fig.
- The flowers are pollinated by the 'fig wasp'. The young figs are green and they turn red or orange on maturing. Each fig is 2-3 cm wide, with a short stalk and is somewhat spherical with a small opening at the apex. The fruits are found inside the figs. The figs are eaten by herbivores, specially birds like koel, hornbill, mynas and bulbuls. It is also cherished by bats, butterflies and monkeys.

Names

The Cluster Fig is known as 'Goolar' in Hindi, 'Umber' in Sanskrit. Botanically it is called 'Ficus recemosa' or 'Ficus glomerata'. It belongs to the Family Moraceae.

Uses

The Cluster Fig is an important medicinal plant. The fig is used in traditional medicines. The figs are carminative while the milky latex is used to treat piles and diarrhea. The leaves are excellent fodder for many herbivores like goats, sheep and elephants. An astringent lotion is prepared from the bark to treat deep wounds. The dried leaves are powered and used to cure biliousness.

The figs are eaten by forest-dwellers. The wood is used for sacrificial fires like havans and yagnas. The tree is considered as sacred.

Ecological Importance

The Cluster Fig plays a very important role in the terrestrial ecosystem. Its leaves and fruits provide food for a variety of herbivores. More than a dozen birds are fond of its ripe figs. At the time of fruiting it becomes like a paradise for birds.

Its pollination by a specific wasp is a very good example of mutualistic relationship found in nature. The fig and the fig wasp seem to be made for each other, because in the absence of one the lifecycle of the other cannot be completed. The leaves of the Cluster Fig are a good food for the larva of the Common Emigrant Butterfly, the Common Crow Butterfly and the Common Map Butterfly. In all, it is a very important tree in our surroundings.

1.8.8 Field Study of Simple Ecosystems: Pond, River and Hill Slopes

Guidelines

- Define ecosystem and its typical features. Identify degradation of ecosystem. - How degradation can be prevented.
- Observe - water cycle, carbon cycle, oxygen cycle, nitrogen cycle, energy cycle.
- Identify common plants and animals observed.

Forest Ecosystem

- Classify forest type,
- Abiotic and biotic,
- Understand food chains and food pyramid,
- Structure of forest,
- Biomass distribution,
- Temperature, moisture of forest,
- Species diversity and animal species,
- Water cycle,
- Feeding pattern of animals..

Grassland Ecosystem

- Seasonal changes of grassland,
- Abundance of grassland species,
- Bird watching,
- Insect world,
- Food chain of grassland.

Desert

- Pattern and structure,
- Birds and insects present,
- Observe typical species,
- Mammals observed.

Aquat

Pattern and structure and nature of ecosystem,

- Observe vegetation, food chain and food pyramid,
- Ecosystem,
- Diversity,
- Effects of variation in season,
- Observe species, water cycle.

Chapter 2

Environmental Pollution

2.1 Introduction to Pollution

Pollution can be defined as the presence of a substance which has harmful or poisonous effects.

Causes

Pollution occurs in different forms such as water, air, radioactive, soil, noise, heat or thermal and light. Every form of the pollution has two sources of occurrence. The point and the non-point sources.

The point sources are easy to identify, monitor and control, whereas the nonpoint sources are hard to control. According to the World Health Organization (WHO), more than 1.1 billion people like in urban areas where the outdoor air is unhealthy to breath.

Air Pollution

The presence of one or more contaminants like smoke, dust, mist and odor in the atmosphere which are injurious to human beings, plants and animals.

Common Air Pollutants Sources and their Effects

According to the World Health Organization (WHO), more than 1.1 billion people live in urban areas where outdoor air is unhealthy to breath. Some of the common air pollutants are as follows:

Carbon Monoxide (CO)

It is a colorless gas that is poisonous to air breathing animals. It is usually formed during the incomplete combustion of carbon containing fuels.

$$2C + O_2 \rightarrow 2CO$$

- Human Source: Cigarette smoking, incomplete burning of fossil fuels. About 77% comes from motor vehicle exhaust.
- Health Effects: Reacts with hemoglobin in red blood cells and reduces the ability to bring oxygen to body cells and tissues by the blood. Which can

cause anemia and headaches. At high levels it leads to coma, irreversible brain cell damage and death.

Nitrogen Dioxide (NO_2)

It is a reddish-brown irritating gas that gives the photochemical smog. In the atmosphere it can be converted into nitric acid.

$$NO_2 + \text{Moisture} \rightarrow HNO_3$$

- Human Source: Fossil fuel burning in motor vehicles (49%) and power industrial plants (49%).
- Health Effects: Lungs irritation and damage.
- Environmental effects: Acid deposition of HNO_3 can damage trees, soils and aquatic life in lakes, HNO_3 can corrode metals and eat away stone on buildings, statues and monuments. NO_2 can damage fabrics.

Sulfur Dioxide (SO_2)

It is a colorless and irritating gas. It is formed mostly from the combustion of sulfur containing fossil fuels such as coal and oil. In the atmosphere, it can be converted to sulfuric acid which is a major component of acid deposition.

- Human Sources: Coal burning in power plants (88%) and industrial processes (10%).
- Health Effects: Breathing problems for healthy people.
- Environmental effects: Reduce visibility, acid deposition of H_2So_4 can damage trees, soils and aquatic life in lakes.

Hydrocarbons

Hydrocarbons especially lower hydrocarbons gets accumulator due to the decay of vegetable matter.

- Human source: Agriculture, decay of plants and burning of wet logs.
- Health effects: Carcinogenic.

Ozone (O_3)

Highly reactive irritating gas with an unpleasant odor that forms in the troposphere. It is a major component of the photochemical smog.

- Human Source: Chemical reaction with volatile organic compounds and nitrogen oxides.

- Environmental Effects: Moderates the climate.

Chromium (Cr)

It is solid toxic metal, emitted into the atmosphere as particulate matter.

- Human Source: Paint, Smelters, Chromium plating.
- Health Effects: Perforation of nasal septum, chrome holes.

Control Measures

The atmosphere has several built-in self-cleaning processes such as dispersion, gravitational settling, flocculation, absorption, rain washout and so on, to cleanse the atmosphere.

In terms of a long range of air pollution, control of contaminants at their source is a more desirable and effective method through preventive or control or technologies.

Source Control

Since we know the substances that causes air pollution, the first approach to its control will be through source reduction:

- Use only unleaded petrol.
- Use petroleum products and other fuels that have low sulfur and ash content.
- Reduces the number of private vehicles on the road by developing an effective public transport system.
- Ensure that houses, schools, restaurants and places where children play are not located on busy streets.
- Plant trees along busy streets because they remove particulates and carbon monoxide and absorb noise.
- Industries and waste disposal sites should be situated outside the city center preferably downwind of the city.
- Use catalytic converters to help control the emission of carbon monoxide and hydrocarbons.

Control Measures in Industrial Centers

- The emission rates should be restricted to permissible levels by each and every industry.
- Continuous monitoring of the atmosphere for the pollutants should be carried out to know the emission levels.

- Incorporation of air pollution control equipment's in the design of the plant layout must be made mandatory.

Public Health Aspects

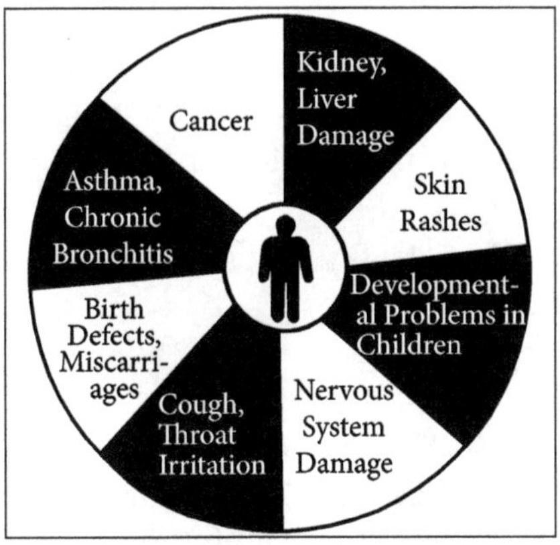

Effects on public health.

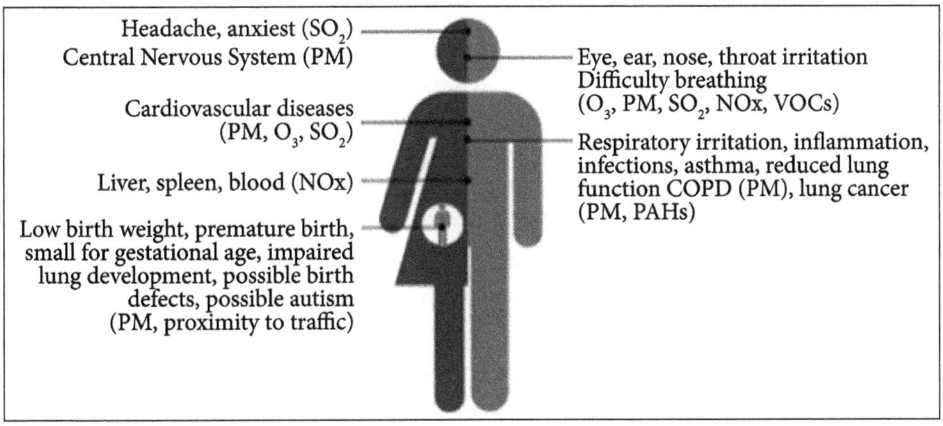

Health Effects of Air pollution.

The Air Pollution and Respiratory Health Branch leads CDC's fight against the environmental related respiratory illnesses which is including asthma and studies the indoor and outdoor air pollution.

Air Pollution and Repository Health

- Asthma: Asthma is a serious environmental health threat, but it can be controlled by taking the medication and by avoiding contact with the environmental "triggers" such as furry pets, dust mites, tobacco smoke, mold and certain chemicals.

- Mold: Exposure to damp and moldy environments may cause throat irritation, nasal stuffiness, eye irritation, coughing or wheezing or skin irritation.

- Carbon Monoxide Poisoning: Carbon monoxide (CO), which is an odorless, colorless gas that can cause sudden illness and death, is found in combustion fumes produced by generators, cars and trucks, lanterns, stoves, gas ranges, burning charcoal and wood and heating systems.

Climate and Health

Climate and Health Program works to prevent and adapt to the health impacts of the extreme weather and the other climate related issues.

Environmental Public Health Tracking

Environmental Public Health Tracking is the ongoing collection, the integration, analysis, interpretation and dissemination of data on environmental hazards, exposures to those hazards and health effects which may be related to the exposures. The goal of tracking is to provide the information which can be used to plan, apply and evaluate the actions to prevent and control the environmentally related diseases.

Health Studies: The Health Studies Branch investigates the human health effects of exposure to environmental hazards ranging from chemical pollutants to natural, technologic or terrorist disasters. The results are used to develop, implement and evaluate the strategies which can help in preventing or reducing the harmful exposures.

Radiation Studies: The Radiation Studies Branch identifies the potentially harmful environmental exposures to ionizing radiation and associated toxicants, conducts the energy related health research and responds to protect public's health in the event of an emergency involving radiation or radioactive materials.

2.1.1 Atmospheric Chemistry: Chemical Composition of the Atmosphere

For convenience of study, atmospheric scientists divide the atmosphere as if it consists of 4 layers. The division is mainly due to temperature variations as the altitude increases. The 4 layers according to the variation of temperature are as follows:

- Ionosphere (Aurora) or Thermosphere.
- Mesosphere.
- Stratosphere.
- Troposphere.

Above 100 km is the thermosphere and ionosphere where the temperature increases from 200 K at 100 km to 500 K at 300 km. The temperature goes even higher as the altitude increases. activity as the altitude decrease. In the outer space, most particles consist of single atoms such as H, He and O etc. At lower altitude (200 - 100 km), diatomic molecules N_2, O_2, NO etc are present. The ionosphere is full of electrically charged ions. The UV rays ionizes these gases.

In the ionosphere:

$$O + h\nu \rightarrow O^{+} + e$$

$$N + h\nu \rightarrow N^{+} + e$$

In the neutral thermosphere:

$$N + O_2 \rightarrow NO + O$$

$$N + NO \rightarrow N_2 + O$$

$$O + O \rightarrow O_2$$

Beyond the neutral thermosphere is the ionosphere and exosphere. These layers are of course interesting for space explorations and environmental concerns and space sciences. The atmosphere in the outer space is more like a plasma than a gas.

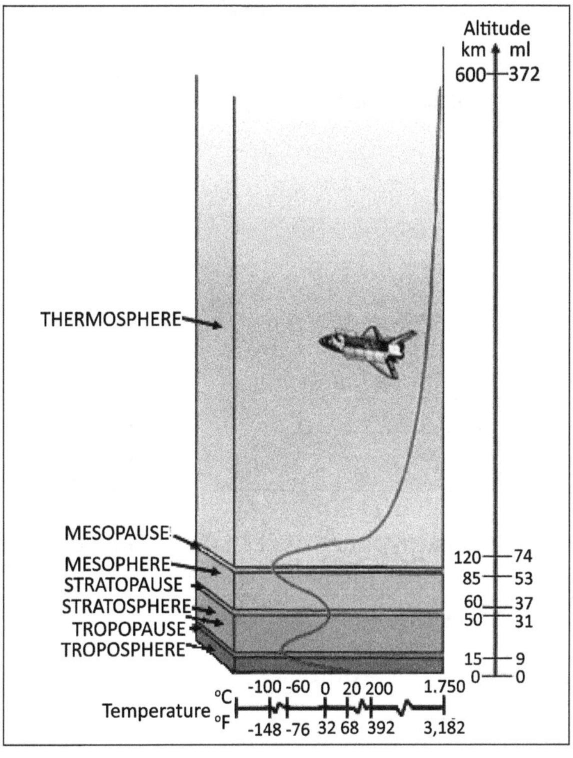

Below the thermosphere is the mesosphere (100 - 50 km) in which the temperature decreases as the altitude increase. In this region, OH, H, NO, HO_2, O_2 and O_3 are common and the most prominent chemical reactions are:

$$H_2O + h\,v \rightarrow OH + HH_2O_2 + O \rightarrow OH + OH$$

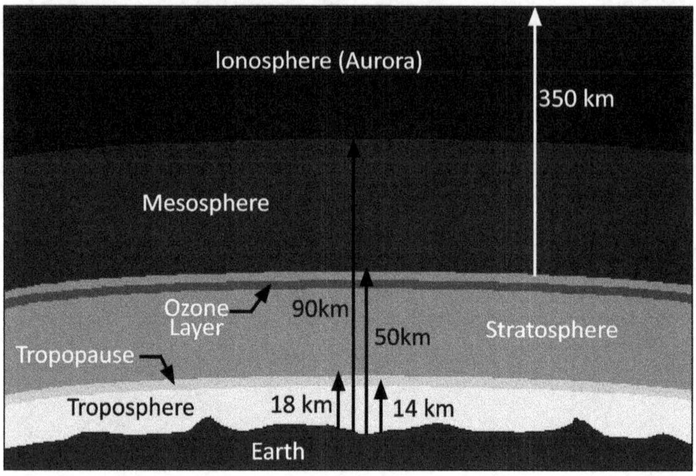

Below the mesosphere is the stratosphere, in which the temperature increases as the altitude increase from 10 km to 50 km. In this region, the following reactions are common:

$$NO_2 \rightarrow NO + ON_2O \rightarrow N_2 + OH_2 + O \rightarrow OH + HCH_4 + O \rightarrow OH + CH_3$$

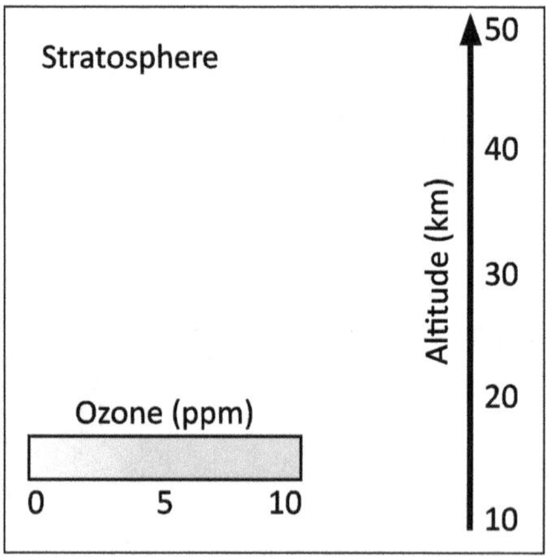

Air flow is horizontal in the stratosphere. A thin ozone layer in the upper stratosphere has a high concentration of ozone. This layer is primarily responsible for absorbing the ultraviolet radiation from the sun.

The ozone is generated by these reactions:

$$O_2 + h\,v \rightarrow O + OO_2 + O \rightarrow O_3$$

The troposphere is the region of rising and falling packets of air. The air pressure at the top of the troposphere is only 10% of that at sea level. There is a thin buffer zone between the troposphere and the next layer called the tropopause.

The major components in the region close to the surface of the Earth are N2 (78%), O2 (21%), Ar (1%) with variable amounts of CO_2, H_2O, NO_2, CH_4, CO, N_2O and O_3. The ozone concentration in this layer is low, about 8% of the total ozone in the atmosphere is in the troposphere. All the weather phenomena happen in the troposphere.

Pollutants in the Atmosphere

For identification purposes, we need to identify the gases produced by man-made process. Some of the gases due to human activities are as follows:

- Carbon monoxide produced by automobiles. This odourless and colorless gas is very toxic.
- Carbon dioxide result from the excess burning of carbon-containing fuel.
- Nitrogen oxides such as NO, NO_2, N_2O_4 due to the production of NO in the internal combustion engine.
- Ozone produced in the exhaust of the internal combustion engine and the variation of ozone concentration in the stratosphere.
- Sulfur oxides produced in mining operation and in the combustion of sulfur containing fuel. Sulfur oxide causes the so called acid rain problem.
- Methane gas produced due to treatments of large amount of waste.
- Chlorofluorocarbons (CFC) are the gases used as refrigerant. When disposed into the atmosphere, they cause the ozone concentration to decrease.
- Water vapour is also considered a greenhouse gas, but it is also generated by nature continuously due to radiation from the Sun. Of course, when water vapour condense into a liquid, much energy is released in the exothermal process. The condensation of water vapour causes storms and many of the weather phenomena.

2.1.2 Chemical and Photochemical Reactions in the Atmosphere: Formation of Smog

Photochemical smog is formed by the reactions of automobile exhaust in the presence of sunlight. Reactions initiated by photons are called photochemical reactions. Smog begins with certain primary pollutants. These substances may or may not be objectionable by themselves. These are transformed by sunlight or by ordinary chemical reactions into the secondary pollutants. The secondary pollutants are involved directly in the buildup of smog. Nitric oxide is a good example of a primary pollutant. It is formed at the high temperatures inside an internal combustion engine when nitrogen and oxygen from air react. Nitric oxide does not build up in the air because it is rapidly converted to nitrogen dioxide.

$$N_2(g)+O_2(g)\rightarrow 2NO(g)$$

$$2NO(g)+O_2(g)\rightarrow 2NO_2(g)$$

Nitrogen dioxide is involved in a chain of reactions that produce ozone. First NO2 is photo chemically decomposed by sunlight.

$$NO_2(g)+hn\rightarrow NO(g)+O(g)$$

Oxygen atoms initiate a number of reactions in polluted air. An important one is the formation of ozone.

$$O(g)+O_2(g)\rightarrow O_3(g)$$

Exposure to 0.1 to 1.0 ppm of ozone produces headaches, burning eyes and irritation to the respiratory passages. Another secondary pollutant is peroxyacetyl nitrate, which is also known as PAN. PAN literally brings tears to our eyes. It is an example of a lachrymator, a compound that causes burning of the eyes and tears. Onions, as we know, contain a lachrymator.

Unburned hydrocarbons in automobile exhaust also produce secondary air pollutants. The oxidation of unburned hydrocarbons produces various alcohols and organic acids which is condense to produce an aerosol. Aerosols are liquid droplets dispersed in the air. They are objectionable because they reduce visibility and make the air look hazy.

Efforts to control smog are usually focused on reducing the source of primary pollutants. The catalytic converters on automobiles are designed to remove the NO, CO and unburned hydrocarbons from automobile exhaust. The catalyst contains platinum and palladium metals. Nitrogen monoxide, a primary pollutant, is converted back to nitrogen and oxygen by a reaction whose rate is increased by the catalyst in the converter.

$$2NO(g)\rightarrow N_2(g)+O_2(g)$$

Carbon monoxide and unburned hydrocarbons are both oxidized to carbon dioxide by the catalytic converter, as shown by the following equations (C_5H_{12} represents a typical hydrocarbon).

$$2CO(g)+O_2(g)\rightarrow 2CO_2(g)C_5H_{12}(g)+8O_2(g)\rightarrow 5CO_2(g)+6H_2O(l)$$

2.1.3 PAN ($C_2H_3O_5N$)

PAN (Peroxyacytyl nitrate) is a kind of air pollution. It is part of smog. PAN makes people's eyes hurt and it is bad for our lungs. It also damages plants.

PAN forms when some other kinds of chemicals get together in the air. One of the chemicals that makes PAN is nitrogen dioxide (NO_2). Nitrogen dioxide comes from the engines of cars and trucks and from factories and power plants that burn coal. Other chemicals that help make PAN are called Volatile Organic Compounds (VOCs). VOCs get into the air from paint, gasoline and other chemicals. Sunlight changes VOCs into other chemicals. Those chemicals get together with oxygen and nitrogen dioxide to make PAN.

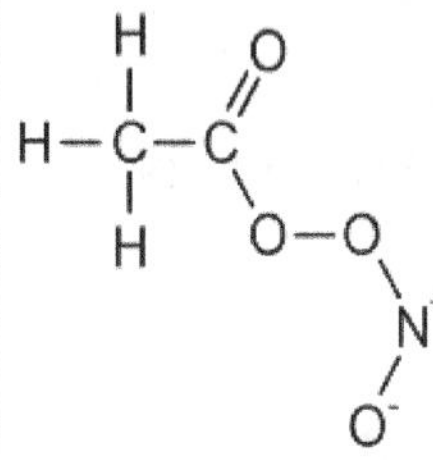

$C_2H_3O_5N$

2.1.4 Acid Rain

Acid rain is one of the most dangerous and widespread forms of pollution.

Acid rain.

Sometimes called “the unseen plague”, acid rain can go undetected in an area for years. Technically, acid rain is a rain that has larger amount of acid in it than normal.

Acid rain is caused by smoke and gases that are given off by factories and cars that run on fossil fuels. When these fuels are burned to produce energy, the sulfur that is present in fuel combines with oxygen and becomes sulfur dioxide. Some of the nitrogen in the air becomes nitrogen oxide. They go into the atmosphere and become acid.

Acid rain is an extremely destructive form of pollution and the environment suffers from its effects. Forests, trees, lakes, animals and plants suffer from acid rain. Humans will become seriously ill and can even die from the effects of acid rain. One of the major problems that acid rain can cause is respiratory problems in human being. Many can even find it difficult to breathe, especially people who have asthma. Asthma, along with dry coughs, headaches and throat irritations can also be caused by the sulfur dioxides and nitrogen oxides from acid rain.

Acid rain can be absorbed by both animals and plants. When the humans eat these plants or animals, toxins inside of their meals can affect them. Kidney problems, brain damage and Alzheimer's disease has been linked to people eating the "toxic" animals/ plants.

Trees are an extremely important natural resource. They regulate local climate, provide timber and forests are homes to wildlife. Acid rain can make trees lose their leaves or needles. The needles and leaves of the trees turn brown and fall off.

Effect of acid rain on forest.

Trees can also suffer from the stunted growth and have damaged bark and leaves, which makes them very much vulnerable to weather, disease and insects. All of this happens partly because of direct contact between the trees and acid rain, but it also happens when the trees absorbs soil that has come into contact with acid rain. The soil poisons the tree with toxic substances such that rain has deposited into it.

Lakes are also damaged by acid rain. A lake polluted by acid rain will support only the hardiest species. Fishes die off and that removes the main source of food for the birds. Also, birds can die from eating the "toxic" fish and insects. Just as birds can be killed from eating toxic fish, fish can die from eating animals that are toxic.

Acid rain can even kill the fish before they are born. Acid rain hits the lakes mostly in the springtime, when fish lay their eggs. The eggs come into contact with acid and the entire generation can be killed. Fish generally die only when the acid level of a lake is high. When the acid level is lower, they can become sick, suffer stunted growth or lose their ability to reproduce.

Architecture and artwork can be destroyed by acid rain. Acid particles can land on buildings, causing corrosion. When sulfur pollutants fall on the surfaces of buildings, they react with the minerals in the stone to form a powdery substance that can be washed away by rain. This powdery substance is known as gypsum. Acid rain can damage buildings, stained glass, railroad lines, airplanes, cars, steel bridges and underground pipes.

Castle stonework in the UK damaged by acid rain.

The below picture shows how acid rain has eroded the stonework of a castle in Lincoln-shire, England and has left a figurehead barely visible.

Stonework of a castle in Lincoln-shire, England damaged by acid rain.

Currently, both airplane industry and railway industry have to spend a lot of money to repair the corrosive damage done by acid rain. Also, bridges have collapsed in the past due to acid rain corrosion.

2.1.5 Oxygen and Ozone Chemistry

Ozone Chemistry

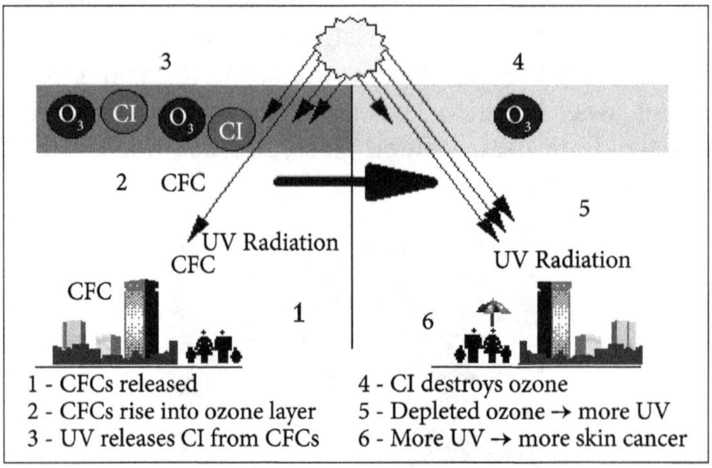

The process of ozone depletion begins when CFCs and other ozone-depleting substances (ODS) are emitted into the atmosphere. Winds efficiently mix the troposphere and evenly distribute the gases. CFCs are extremely stable and they do not dissolve in the rain. After a period of several years, an ODS molecules reach that the stratosphere about 10 kilometers above the Earth's surface.

Strong UV light breaks the ODS molecule. CFCs, HCFCs, carbon tetrachloride, methyl chloroform and other gases releases the chlorine atoms and halons and methyl bromide releases the bromine atoms. It is these atoms that destroy ozone, not the intact ODS molecule. It is estimated about one chlorine atom can destroy about 100,000 ozone molecules before it is removed from the stratosphere.

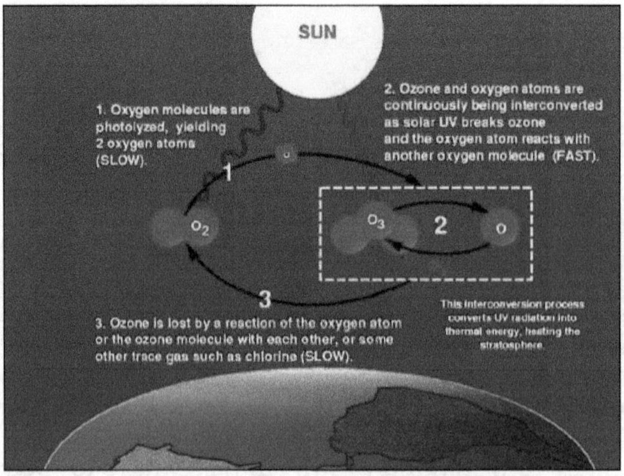

Ozone is constantly produced and destroyed in a natural cycle, as shown in the figure above, courtesy of NASA GSFC. However, the overall amount of ozone is essentially stable. This balance can be seen of as a stream's depth at a particular location. Although individual water molecules are moving past the observer, the total depth remains constant. While ozone production and destruction are balanced at ozone levels remain stable. This was the situation until the past several decades of large, increases in the stratospheric chlorine and bromine, however, have upset that balance. In effect, they have added a siphon downstream, removing the ozone faster than natural ozone creation reactions can keep up. Therefore, ozone levels fall.

Since ozone filters out harmful UVB radiation, less ozone means higher UVB levels at the surface, the larger increase in incoming UVB. UVB has been linked to skin cancer, cataracts, damage to materials like plastics and harm to certain crops and marine organisms. Although some UVB reaches the surface even without ozone depletion, its harmful effects will increase as a result of this problem.

2.1.6 Mitigation Procedures: Control of Particulate and Gaseous Emission

Airborne particulate matters (PM) emissions can be minimized by pollution prevention and emission control measures. Prevention, which is frequently more cost-effective than control, should be emphasized. Special attention should be given to mitigate the effects, where toxics associated with the particulate emissions may pose a significant environmental risk.

Measures such as improved process design, operation, maintenance, housekeeping and other management practices can reduce emissions. By improving the combustion efficiency in Diesel engines, generation of particulate matters can be significantly reduced. Proper fuel-firing practices and combustion zone configuration, along with the adequate amount of excess air, can achieve lower PICs.

Few following steps should be adhered to control PM:

- Low-ash fossil fuels contain less non-combustible, ash-forming mineral matter and thus generates the lower levels of particulate emissions.
- Choosing the cleaner fuels. Natural gas used as fuel emits the negligible amounts of particulate matter.
- Use of more efficient technologies or process changes can reduce the PIC emissions.
- Reduction of ash by coal cleaning reduces the generation of Particulate Matter (PM) emissions.
- A variety of particulate removal technologies are as follows:
    - Electrostatic precipitators (ESPs).

  - Inertial or impingement separators.
  - Wet scrubbers that rely on a liquid spray to remove the dust particles from a gas stream.
  - Filters and dust collectors.
- Advanced coal combustion technologies such as coal gasification and fluidized-bed combustion are examples of cleaner processes that may lower PICs by approximately 10%.

The principal methods for controlling the release of particulate matter are as follows:

- Consider the alternative fuels such as gas instead of coal.
- Identify the measures for improving the operating and management practices.
- Consider the alternative production processes and technologies, such as fluidized bed combustion, that results in the reduced PM emissions.
- Consider the fuel-cleaning options such as coal washing, which can reduce ash content by up to 40%.
- Select the optimal particulate removal devices such as ESPs and Bughouses.

2.1.7 Control of SO_2, NO_X, CO and HC

Estimation of SO_2

Emissions from Fuel Composition If the sulphur content is known, SO, emissions can be easily calculated from fuel consumption data. For example, if a fuel can 0.15% sulphur. 20 kg of fuel releases 60 g of SO_2. This calculation presupposes that all the sulphur a bung according to stoichiometric relationship and no flue gas dcsulphurization equipment is installed.

Estimation of NOx, HC and Heavy Metals

Usually, Estimation of NO and HC emissions would not be in agreement with stoichiometric calculations. The basic reason is that actual NO, emissions are dependent upon the furnace temperature and ingress of atmospheric nitrogen dining burning. Thus, actual emission of NO would be higher than the theoretically calculated quantities.

On the other band, hydrocarbon emissions would be lower than the calculated values owing to incomplete burning in the engine. It is well known that thermodynamically 100% combustion of the fuel is impossible in IC engines. Heavy metal emissions can be calculated to a fair degree of accuracy in heavy oils, coal, naphtha, etc. Light oils contain very low levels of heavy metals.

2.2 Water Pollution: Physical and Chemical Properties of Terrestrial and Marine Water

Water is essential to the maintenance of all life. It constitutes 80% or more by weight of active protoplasm. It is the most efficient of all solvents and carries in solution the necessary gases, oxygen and carbon dioxide, as well as the mineral substances necessary to the growth of plants and animals and it is itself one of the essential raw materials in the manufacture of foods by plants.

Organisms living in the terrestrial environment have devised means, such as impervious integuments, to conserve water and the land plants have roots and special vascular systems for transport of water to all growing parts. In the marine environment, there is freedom from dessication, except at the high-tide levels and therefore no highly specialized means are provided for conservation of water or for its transport in plants. Also of biological importance are the high heat capacity of water and its high latent heat of evaporation in which both obviate the danger that might results from rapid change of temperature in the environmental medium.

Owing to the high degree of transparency of water, it is possible for the sea to sustain the plant life throughout a relatively deep layer and in animals, the development of organs of vision and orientation has progressed to a marked degree. Sea water is a buffered solution, i.e., changes from acid to alkaline condition or vice versa, are resisted.

This property is of vital importance to the marine organisms, mainly for two reasons:

- The abundant supply of carbon can be available in the form of carbon dioxide for the use of plants in the synthesis of carbohydrates without disturbance to the animal life that may be sensitive to small changes in pH.
- In the slightly alkaline habitat, many organisms that construct shells of calcium carbonate can carry on this function much more efficiently than in a neutral solution.

The support offered to the bodies of marine organisms by the specific gravity of the surrounding medium obviates the need of special supporting skeletal structure in many forms. Striking examples of these are the jelly fishes, unarmored molluscs, unarmored din flagellates and even the large marine mammals with their heavy skeletons, which could not survive in their present bulky state except in the aquatic habitat. The hard shells of clams, crabs, snails and so on, doubtless serve as support, especially in some burrowing and intertidal forms, but these hard parts may be looked upon also as the protective and framework for the attachment of muscles used in creeping, digging or swimming.

Water pollution is the pollution caused due to any chemical, physical or biological change in water quality that has a harmful effects on living organisms or makes water unsuitable for desired uses.

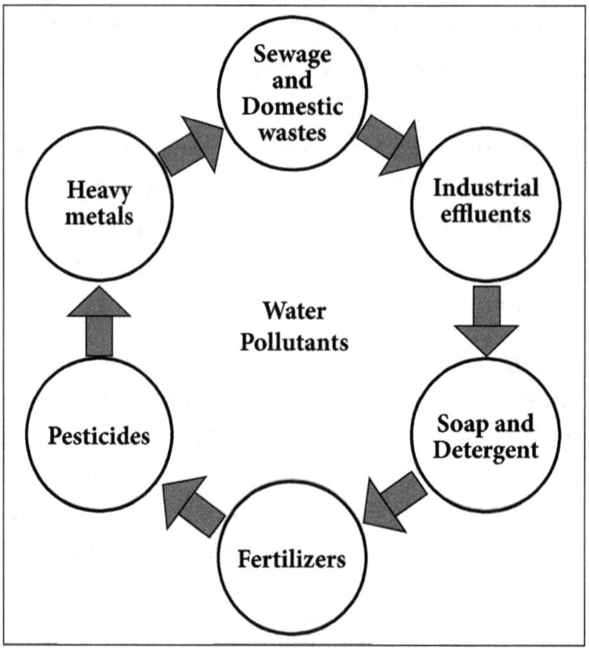

Water pollutants.

Infection Agents:

- Example: Bacteria, Viruses, Protozoa.
- Human Sources: Human and animal wastes.
- Effects: Variety of diseases.

Oxygen Demanding Wastes:

- Example: Organic wastes such as animal manure and plant debris that can be decomposed by aerobic bacteria.
- This degradation consumes dissolved oxygen in water. Do is the amount of oxygen dissolved in a given quantity of water at a particular pressure and temperature.
- The standard point of Do varies from 8 - 15 mg/lit.
- Human Source: Sewage, animal feedlots, paper mills.
- Effects: Large populations of bacteria decomposing these wastes can degrade water quality by depleting water of dissolved oxygen. This causes fish and other forms of oxygen consuming aquatic life to die.

Organic Chemicals:

- Example: Oil, Gasoline, Plastics, Pesticides.

- Human Sources: Industrial effluents, surface runoff from farms, household cleansers.
- Effects:
    - Can threaten human health by causing nervous system damage and some cancers.
    - Harm fish and wildlife.

Sediment:

- Examples: Soil, Silt, etc.
- Human Sources: Land erosion.
- Effects:
    - Disrupt aquatic food webs.
    - Can cloud water and reduce photosynthesis.
    - Settle out and destroy feeding and spawning rounds of fish.
    - Carry pesticides, bacteria and other harmful substances.

Radioactive Materials:

- Example: Radioactive isotopes of iodine, radon, uranium and chromium.
- Human Sources: Nuclear power plants, mining and processing of uranium and other ores, nuclear weapons production and natural sources.
- Effects: Genetic mutations, birth defects and certain cancers.

2.2.1 Water Quality Parameters: Physical, Chemical and Biological

Water quality is determined by physical, chemical and microbiological properties of water. These water quality characteristics throughout the world are characterized with wide variability. Hence, the quality of natural water sources used for different purposes should be established in terms of the specific water-quality parameters that mostly affect the possible use of water.

1. Physical Characteristics of Water

Physical characteristics of water are determined by senses of sight, touch, taste and smell. Example: Temperature by touch, colour, floating debris, turbidity and suspended solids by sight and taste and odour by smell.

Temperature

The temperature of water affects some of the important physical properties and characteristics of water such as thermal capacity, density, specific weight, viscosity, surface tension, specific conductivity, salinity and solubility of dissolved gases and etc. Chemical and biological reaction rates increase with increasing temperature. Reaction rates usually assumed to be double for an increase in temperature of 10°C. The temperature of water in streams and rivers throughout the world varies from 0 to 35°C.

Colour

Colour in water is primarily a concern of water quality for aesthetic reason. Coloured water give the appearance of being unfit to drink, even though the water may be perfectly safe for public use. On the other hand, colour can indicate the presence of organic substances, such as algae or humic compounds. More recently, colour has been used as a quantitative assessment of the presence of potentially hazardous or toxic organic materials in water.

Taste and Odour

Taste and odour are human perceptions of water quality. Human perception of taste includes sour, salty, sweet and bitter. Relatively simple compounds produce sour and salty tastes. However sweet and bitter tastes are produced by more complex organic compounds. Human detect many more tips of odour than tastes. Organic materials discharged directly to water, such as falling leaves, runoff, etc., are sources of tastes and odour-producing compounds released during biodegradation.

Turbidity

Turbidity is a measure of the light-transmitting properties of water and is comprised of the suspended and colloidal material. It is important for health and aesthetic reasons.

Solids

The total solid content of water is defined as the residue remaining after evaporation of the water and drying the residue to a constant weight at 103°C to 105°C. The organic fraction is considered to be related to the loss of weight of the residue remaining after evaporation of the water and after ignition of the residue at the temperature of 500°C.

The volatile solids will oxidize at this temperature and will be driven off as gas. The inorganic remind as inert ash. Solids are classified as settle able solids, suspended solids and filterable solids. Settle able solids are the one that settle under the influence of gravity.

Suspended solids and filterable solids are classified based on the particle size and the retention of suspended solids on the standard glass-fiber filters.

2. Chemical Characteristics of Water

The chemical characteristics of natural water are a reflection of the soils and rocks with which the water has been in contact. In addition, agricultural and urban runoff and municipal and industrial treated wastewater impact the water quality. Microbial and chemical transformations also affect the chemical characteristics of the water.

Inorganic Minerals

Runoff causes erosion and weathering of geological formation, rocks and soils as the runoff travels to the surface-water bodies. During this period of contact with rocks and soils, the water dissolves inorganic minerals, which enter the natural waters.

Major Cations

Major cations found in natural water include calcium (Ca^{2+}), magnesium (Mg^{2+}), sodium (Na^{+}) and potassium (K^{+}). Calcium (Ca^{2+}), is the most prevalent cation in water and second inorganic ion to bicarbonate in most surface water.

The principal concern about calcium is related to the fact that calcium is the primary constituent of water hardness. Calcium precipitates as $CaCO_3$ in iron and steel pipes. A thin layer of $CaCO_3$ can help inhibit corrosion of the metal.

However, excessive accumulation of $CaCO_3$ in boilers, hot water heaters, heat exchangers and associated piping affects heat transfer and could lead to plugging of the piping. Calcium concentration of up to 300 mg/L or higher have been reported. However, calcium concentrations of 40 to 120 mg/L are more common.

Magnesium is not abundant in rocks as calcium. Although magnesium salts are more soluble than calcium, less magnesium is found in surface water. Sodium and potassium are commonly found as free ions. The concentration of these cations in natural water usually are low. Other constituents in natural water in concentration of 1 mg/L or higher include aluminium, boron, iron, manganese, phosphorus and etc.

Major Anions

Major anions include sulfate, chloride, carbonate, bicarbonate, fluoride and nitrate. Bicarbonate (HCO_3^-) is the principal anion found in natural water. These ions are very important in the carbonate system, which provides a buffer capacity to natural water and is responsible in a great measure for the alkalinity of water.

One source of bicarbonate ions (HCO_3^-) in the natural water is the dissociation of carbonic acid (H_2CO_3) that is formed when carbon dioxide (CO_2) from the atmosphere or from animal (e.g. fish) and bacterial respiration, which dissolves in water.

In addition to bicarbonates (HCO_3^-), anions such as chlorides (Cl^-), sulfates (SO_4^{2-}) and nitrates (NO_3^-) are commonly found in natural water. These anions are released during the dissolution and dissociation of common salt deposits in geologic formations.

The concentration of the chlorides anions (Cl^-) determines the water quality because the quality of water get worse after increasing in the concentration of this anions which limit the possibilities of using of natural water for different purposes. Principal source of the chlorides anions (Cl^-) in natural water are magmatic rock formations that include chlorine-content minerals. The second source of this anions is Ward Ocean from where a considerably amount of chlorides anions (Cl^-) enter in the atmosphere. From the atmosphere, chlorides anions (Cl^-) enter in the natural water in the result of interaction between the precipitation and soil.

The sulfates anions (SO_4^{2-}) are frequently found in natural water as the result of the chemical dissolution, dissolve sulfur-content minerals and oxidize sulfates and sulfur.

$$CaCO_3 + H_2SO_4 = CaSO_4 + H_2O + CO_2$$

$$2FeS_2 + 7O_2 + 2H_2O = 2FeSO_4 + 2H_2SO_4$$

$$2S + 3O_2 + 2\,H_2O = 2\,H_2SO_4$$

The sulfates anions (SO_4^{2-}) enter in natural water as the result of the oxidation of the substances from plant and animal origin. The increase concentration of the sulfates anions (SO_4^{2-}), at one hand brings about change for the worse of some physical characteristics of water and on the other hand, has destructive influence upon human consumption. The concentration of the sulfates anions (SO_4^{2-}) fluctuates in a wide range in surface water - from 5 mg/l to 60 mg/l.

Nitrate anions (NO_3) are found in natural water as the result of the bacteriological oxidation of nitrogenous materials in soil. Hence, the concentration of these anions rapidly increases when the process of nitrification takes place very intensively. Another important source for dressing of the surface water with Nitrate anions (NO_3) are precipitations, which absorbs nitric oxides and convert them into nitric acid. A great deal of nitrate anions (NO_3) enter in surface water together with domestic water and water from industry, agriculture and etc. Nitrate anions (NO_3) are one of the indicators for the degree of the pollution with organic nitrate-content substances.

Other anions found in water include fluorides (F^-), carbonates (CO_3^{2-}) and phosphates (PO_4^{3-}). Typical concentrations of major ions in the classic "word average" river are presented in the Table.

Table: Typical concentrations of Major Ions in the Classic "Word Average" River.

Constituent	Concentration mg/L	Cations meg/L	Anions meg/L
	Cations		
Ca^{2+}	15	0.750	-
Mg^{2+}	4.1	0.342	-
Na^{+}	6.3	0.274	-
K^{+}	2.3	0.059	-
	Anions		
HCO_3^-	58.4	-	0.958
SO_4^{2-}	11.2	-	0.233
Cl^-	7.8	-	0.220
NO_3^-	1	-	0.017
Sum	106.1	1.425	1.428

Carbonate Equilibrium

The carbonate - bicarbonate system is presumably the most important chemical system in natural waters. The carbonate system provides the buffering capacity essential for maintaining the pH of natural water systems in the range required by bacteria and other aquatic species.

The carbonate system includes the following species:

$$CO_2,\ H_2CO_3, HCO_3^-,\ CO_3^{2-},\ OH^- \text{ and } H^+.$$

The total content (mol/dm³) of its components is as follows:

$$\sum CO_2 = [CO_2] + [H_2CO_3] + [HCO_{3^-}] + [CO_{3^{2-}}]$$

Frequently Ca^{2+} and $CaCO_3$ are included in the carbonate system, since Ca^{2+} is second in abundance to HCO_3- in natural waters. The solution of carbon dioxide (CO_2) into natural water causes the formation of carbonic acid (H_2CO_3). The H_2CO_3 dissociates to bicarbonate (HCO) and hydrogen (H^+) ions. In its turn HCO_3^- can dissociate and produce carbonate (CO) and hydrogen (H^+) ions.

$$CO_2 + H_2O \rightleftarrows H_2CO_3$$

$$H_2CO_3 \rightleftarrows H^+ + HCO$$

$$HCO_3 \rightleftarrows H^+ + CO$$

The hydrogen ion (H^+) concentrations $[H^+] = 10^{-pH}$ in water controls the pH of the solution. The pH of water is defined as the negative logarithm of the $[H^+]$, where $[H^+]$ is the hydrogen ion concentration expressed in moles per litre (mol/L).

$$pH = -\log\left[H^+\right] = \log\frac{1}{\left[H^+\right]}$$

$$pOH = -\log\left[OH^-\right] = \log\frac{1}{OH^-}$$

The pH is the negative power to which 10 must be raised to equal the hydrogen ion concentration or $[H^+] = 10^{-pH}$. In a neutral solution $[H^+]$ is 10^{-7} or pH = 7. At greater hydrogen ion concentrations the pH is lower and for greater hydroxide ion concentrations the pH increases. The pH range is from 0 (extremely acidic) to 14 (extremely basic).

Therefore, the pH of the water controls which species is predominant. Water molecules HOH (commonly written as H_2O) dissociate or ionize to H^+ and OH^- ions.

$$2H_2O \rightleftharpoons H_3O^+ + OH^-$$

Or written in a simple form:

$$H_2O \rightleftharpoons H^+ + OH^-$$

The product of $[H^+]$ and $[OH^-]$, in mol/L, is constant:

$$\left[H^+\right]\left[OH^-\right] = K = 1\times 10^{-14}$$

Where K is the ion-product constant of water.

If the hydrogen ion concentration is 10^{-4} mol/L the hydroxide ions (OH^-) concentration must be $10^{-14}/10^{-4} = 10^{-10}\,mol/L$. Since the 10^{-10} is smaller than 10^{-4}, the solution is acidic. A large amount of hydrogen ions (H^+) in water makes the water acid and lack of hydrogen ions makes the water basic. A basic solution has predominance of hydroxide ions (OH^-). The dissociation reactions of carbonic acid (H_2CO_3) and of the bicarbonate ion (HCO) can be written as the following equations:

$$H_2CO_3 \rightleftharpoons HCO + H^+ \text{ and } k_1 = 10^{-6.35} \text{ at } 25\ °C;$$

$$H_2CO_3 \rightleftharpoons CO + H^+ \text{and } k_2 = 10^{-10.3} \text{ at } 25\ °C$$

These equations can be used to define the relative distribution of carbonate species as a function of pH. Changes in pH can have drastic effects on the species present in the carbonate system. Therefore the fraction of carbonic acid $[H_2CO_3]$ may be expressed as αo and written as the fraction:

$$\alpha_o = \frac{\left[H_2\,CO_3\right]}{\left[H_2\,CO_3\right]+\left[HCO_3^-\right]+\left[CO_3^{2-}\right]}$$

Similarly the fraction of bicarbonate ion $[HCO_3^-]$ and carbonate ions $[CO_3^{2-}]$ may be expressed as α_1 and α_2, respectively and written as the following fractions:

$$\alpha_1 = \frac{\left[H_2\,CO_{3^-}\right]}{\left[H_2\,CO_3\right]+\left[HCO_3^-\right]+\left[CO_3^{2-}\right]}$$

$$\alpha_2 = \frac{\left[CO_3^{2-}\right]}{\left[H_2\,CO_3\right]+\left[HCO_3^-\right]+\left[CO_3^{2-}\right]}$$

PH and Alkalinity

Alkalinity is defined as the capacity of natural water to neutralize acid added to it. Total alkalinity is the amount of acid required to reach a specific pH (pH = 4,3 to 4,8). Total alkalinity can be approximated by alkalinity as:

Total alkalinity = $\left[HCO_{3^-}\right]+2\left[CO_{3^{2-}}\right]+\left[OH^-\right]-\left[H^+\right]$

Total alkalinity includes Hydroxide alkalinity $[OH^-]$, Bicarbonate alkalinity $[HCO_3^-]$ and Carbonate alkalinity $[CO_3^{2-}]$. If the pH and total alkalinity are measured, the concentration of the various components of alkalinity can be calculated using the values of α_1 and α_2 determined for the pH of the water. These values can be used to calculate the:

Hydroxide alkalinity = $\dfrac{K_W}{\left[H^+\right]}$

Bicarbonate alkalinity = $\alpha_1 C_t$

Carbonate alkalinity = $2\,\alpha_2 C_t$

Where C_t is total carbonate and $C_t = \left[H_2CO_3\right]+\left[HCO_{3^-}\right]+\left[CO_{3^{2-}}\right]$.

The amount of strong acid (in eq/L) required to change colour of the water from pink to clear (colour less) when a small amount of phenolphthalein reagent is put into the water sample is phenolphthalein alkalinity.

This colour change occurs at approximately pH = 8.3. Continuing the titration to pH = 4.3, which is the H_2CO_3 endpoint, yields the total alkalinity. The values of each three forms of alkalinity can be determined using the relative values of the phenolphthalein alkalinity and the total alkalinity, expressed as either eq/L or mg/L $CaCO_3$.

Acidity

Acidity is the "quantitative capacity of aqueous media to react with hydroxyl ions". Titration with a strong base (NaOH) to define end points (pH = 4,3 and pH = 8,3). Acidity indicates the corrosiveness of acidic water on steel, concrete and other materials.

Inorganic Indicators of Water Quality

The inorganic parameters include hardness, total dissolved solids, conductivity and adsorption ratio.

Hardness

Hardness is correlated with TDS (Total dissolved solids). It represents total concentration of Ca^{2+} and Mg^{2+} ions and is reported in equivalent $CaCO_3$. Other ions (Fe^{2+}) may also contribute. Hardness expressed as mg/L $CaCO_3$ is used to classify waters from "soft" to "very hard". This classification is summarized in Table.

Table: Relationship of hardness concentration and classification of natural water.

Hardness as mg/L $CaCO_3$	Classification
0 - 60	Soft
61 - 120	Moderately hard
121 - 180	Hard
>180	Very hard

Hardness observed for streams and rivers throughout the world ranges between 1 to 1000 mg/L. Typical concentrations are 47 mg/L to 74 mg/l CaCO3.

Hardness is an indicator to industry of potential precipitation of calcium carbonates in cooling towers and boilers, interference with soaps and dyes in cleaning and textile industries and with emulsifiers in photographic development. Hard water is less corrosive than soft. Treatment usually left to consumer depending on needs.

Total Dissolved Solids

Total dissolved solids is a measure of salt dissolved in the water sample after the removal of suspended solids. TDS is the residue remaining after the evaporation of the water. The TDS load carried in streams throughout the world has been estimated by Livingston (1963) to 120 mg/L.

Conductivity

The concentration of total dissolved solids is related to electrical conductivity (EC; mhos/cm) or specific conductance. The conductivity measures the capacity of water to transmit electrical current. The conductivity is a relative term and the relationship between the TDS concentration and conductivity is unique to a given water sample and in a specific TDS concentration range. Conductivity increases as the concentration of TDS increases.

TDS and conductivity affects the water sample and the solubility of slightly soluble compounds and gases in water. In general, the corrosiveness of the water increases as TDS and EC increase, assuming other variables are kept constant.

Sodium Adsorption Ratio

The Sodium Adsorption Ratio (SAR) is used to evaluate the hazard in irrigation waters caused by sodium (Na^+). The SAR relates the concentration of sodium ions to the concentration of magnesium (Mg^{2+}) and calcium (Ca^{2+}) ions. The SAR is defined as:

$$SAR = \frac{Na^+}{\sqrt{Ca^{2+} + Mg^{2+}} / 2} \text{ in meg / L}$$

The proper ratio of sodium ions to calcium and magnesium ions in irrigation water results in irrigated soil that is granular in texture, easily worked and permeable. With increasing proportions of sodium as the SAR increases, soil tends to become less permeable and more difficult to work.

Radionuclides

Radionuclides in water are classified according to the type of energy released - alpha radiation, beta radiation or gamma radiation.

Natural radiation is found in elements in the Earth's crust (potassium-40 (^{40}K)). Another source of natural radiation results from cosmic ray bombardment in the atmosphere (tritium (^{3}H) and carbon-14 (^{14}C)). Other high-atomic-weight, naturally occurring isotopes found in natural water include uranium-238, thorium-232, uranium-235 and breakdown products as radium-226 and radium-228.

The units of radiation measurements are curies (C_I) or rems (C_I = 3,7 X 10^{10} nuclear transformations per second; picocurie (pC_I) = 10^{-12} C_I). A rem is the radiation dose producing the same biological effect (rem = absorbed dose X quality factors).

Each type of radiation has different health effects. For example, alpha particles travel at velocities up to 10^7 m/sec. When ingested the relatively massive alpha particles can be very damaging to body tissue. Beta particles travel at about the speed of light, penetrate to greater depth because of their smaller mass and create less damage. Gamma radiation penetrates deeply, but has limited effects at low levels. The body dose that accurses from the drinking water compared to the natural background radiation is low. However, EPA policy assumes that potential harm exists from any level of radiation.

Organic Materials

Organic chemicals are made up of carbon (C), hydrogen (H), as well as nitrogen (N) and oxygen (O). Organic compounds are derived from living organism as well as industrial sources. A wide variety of assortments of organic compounds are produced in the chemical and petrochemical industries. The organic compounds also may contain phosphorus (P), sulfur (S), chlorine (Cl), fluorine (F), bromine (Br) and iodine (I).

Organic compounds in water also affect the water quality. Organic chemicals cause disagreeable tastes and odours in drinking water. Vinyl chloride, benzene and other organic contaminants are known carcinogenic agents, while chloroform is a cancer-suspect agent.

Natural Organic Matter

Organic materials are found in natural water as a result of wide range of processes, together with the precipitation and surface water, as a result of the interaction between soils and precipitation and etc. Organic materials in soils originate from plant and animal degradation products. These degradation products condense and polymerize into fulvic and humic acids to kerogen and finally coal.

Chemical and microbial processes cause the transformation by first attacking functional groups and aliphatic side chains. Condensation and polymerization of various reactive groups results in larger, more aromatic molecules that decrease in solubility until kerogen is produced. The end products are not soluble in acid or alkali and are resistant to biodegradation and chemical reaction.

Man-Made Organics

Synthetic organic compounds includes a broad variety of aliphatic and aromatic compounds. Many manufactured organic compounds may be found at very low concentrations in natural water. Isolation, identification and evaluation of health effects of these synthetic organics at low concentrations are lacking.

Measurement of Organics in Water: Organic Carbon

Organics in water can be expressed in terms total organic carbon (TOC). The TOC is the difference between total carbon (TC) and inorganic carbon (IC). Typical concentrations of organic matter observed in natural water in streams and rivers throughout the world are presented in the below Table.

Table: Water quality of streams and rivers and organic materials.

Water quality parameter	Typical value, mg/L	Observed ranges, mg/L
Inorganic carbon (IC)	50	5-250
Total organic carbon (TOC)	1-10	0.01 - 40
Dissolved organic carbon (DOC)	1-6	0.3-32
Volatile organic carbon (VOC)	0.05	
Total organic matter	2-20	0.02 - 80

Organic Indicators of Water Quality

For quantity assessment of concentrations of organic materials, indicator, Total Oxygen Demand are used. Total Oxygen Demand includes Chemical Oxygen Demand, Biochemical Oxygen Demand and Nitrogenous Biochemical Oxygen Demand and can be expressed as:

$$CaH_bO_cN_dS_e + xO_2 \rightarrow aCO_2 + \tfrac{1}{2}\, bH_2O + dNO_{3^-} + eSO_{4^-}$$

Biological Oxygen Demand (Animation)

Biological oxygen demand (BOD), the most widely used parameter, is a measure of the amount of oxygen used by indigenous microbial population in water in response to the introduction of degradable organic material. This parameter depends on water characteristics such as dilution, essential nutrients (N, P, K, Fe, etc) and bacteria seed.

The 5-day BOD (BOD_5) is most widely used. The BOD_5 of natural water is related to the dissolved oxygen concentration, which is measured at zero time and after 5 days of incubation at 20°C. The difference is the dissolved oxygen used by the microorganisms in the biochemical oxidation of organic matter. The BOD5 can be calculated as $BOD_5 = D_0 - D_1$, in which the BOD_5 is in mg/L and D_0 and D_1 are the dissolved oxygen concentration in mg/L at time 0 and 5 days, respectively. Typical concentration of BOD_5 for streams and rivers throughout the world are < 2 to 15 mg/ L and the observed range is < 2 to 65 mg/L.

Chemical Oxygen Demand

The chemical oxygen demand (COD) test of natural water yields the oxygen equivalent of the organic matter which can be oxidized by strong chemical oxidizing agent in the acidic medium. Potassium permanganate is the oxidizing chemical. Silver sulfate is added as a catalyst and to minimize the interference of chloride on the COD test. Mercuric sulfate is also added to inhibit the interferences of metals on the oxidation of organic compounds. The reaction of dichromate with organic matter is expressed as:

$$\text{Organic matter } (C_aH_bO_c) + Cr_2O_{7}^{2-} + H^- \rightarrow 2Cr_{3}^{+} + CO_2 + H_2O$$

The COD observed in natural streams and rivers is < 2 mg/L to 100 mg/L.

Dissolved Gases

The principal transfer of gas in natural water is the transfer of oxygen from the atmosphere to the water. However, gas transfer is also used to strip hydrogen sulfide (H_2S), ammonia (NH_3) and volatile organic compounds (VOC) from the water. In both processes material is transferred from one bulk phase to another across a gas-liquid interface. For example, oxygen is transferred from the bulk gaseous phase across the gas-liquid interface into bulk liquid phase. In the case of striping volatile organic compounds (VOC) from liquid, the VOC is transferred from the bulk liquid phase (water) across the liquid-gas interface into the bulk gaseous phase.

Solubility of Gases

The equilibrium of each phase, concentration of gases or volatile organic compounds dissolved in water, depends on the temperature, type of gas or volatile compounds and partial pressure of the gas or volatile compounds adjacent to the water. The relationship between the partial pressure of the gas in the atmosphere above the water and the concentration of the gas or volatile compound in the water is described by Henry's law:

$$X_g = H\,P_g$$

Where,

P_g = Partial pressure of gas, atm.

H = Henry's law constant

X_g =Equilibrium mole fraction of dissolved gas, mol/dm3

The Henry's law constant is a function of type of gas or volatile organic compounds, temperature of the bulk liquid and constituent of the liquid (water). Values of Henry's constant for various gases that are slightly soluble in water at 20°C are shown in the below table. Thermodynamic parameters that are required to adjust Henry's law constants for different temperatures are also included in the below table.

Table: Henry's law constants for selected gases that is slightly soluble in water.

Gas	Henry's constant at 20°C, atm	Temperature dependence	
		ΔH, kcal/kmol · 10^{-3}	K
Oxygen (O_2)	4.3	1.45	7.11
Nitrogen (N_2)	8.6	1.12	6.85
Methane (CH_4)	3.8	1.54	7.22
Ozone (O_3)	5.0	2.52	8.05
Carbon dioxide (CO_2)	1.51	2.07	6.73
Hydrogen sulfide (H_2S)	5.15	1.85	5.88
Chlorine (Cl_2)	5.85	1.74	5.75
Chlorine dioxide (CLO_2)	54	2.93	6.76
Sulfur dioxide (SO_2)	38	2.40	5.68
Ammonia (NH_3)	0.76	3.75	6.31

Dissolved Oxygen

Typical dissolved oxygen concentrations observed in streams and rivers throughout the world are 3 to 9 mg/L. The observed range of dissolved oxygen concentrations is 0 mg/L to 19 mg/L.

Dissolved oxygen is important in natural water because many microorganisms and fish require it in aquatic system. Dissolved oxygen also establishes an aerobic environment in which the oxidized forms of many constituents in water are predominant. Under anoxic conditions in water, reduced forms of chemical species are formed and frequently lead to the release of undesirable odours until oxic conditions develop.

Biological Characteristics

In a typical aquatic ecosystem, plant and animal materials are composed of carbon, hydrogen, oxygen, nitrogen, phosphorus and sulfur. These elements are building blocks for carbohydrates, lipids, proteins, phospholipids and nucleic acid.

Protein and nucleic acids consist of nitrogen, which is required by organisms in greatest quantity after carbon and oxygen. Organic nitrogen ammonia (NH_3), nitrite (NO_2^-), nitrate (NO_3^-) and nitrogen gas (N_2) are important nitrogen-containing compounds in aquatic systems. The atmosphere is the reservoir for nitrogen.

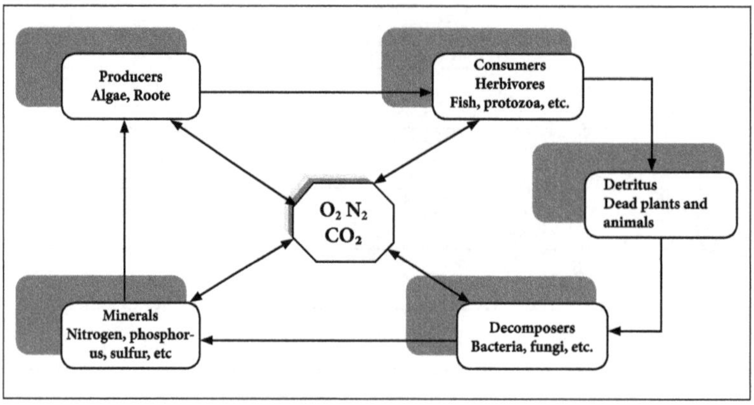

Schematic aquatic ecosystem.

Nitric oxide (NO) is formed during fuel combustion at high temperatures:

$$N_2 + O_2 \rightarrow 2NO$$

NO reacts easily with ozone and peroxides in the atmosphere:

$$NO + O_3 \rightarrow NO_2 + O_2$$

$$NO + H_2O_2 \rightarrow NO_2 + 2OH^-$$

NO_2 reacts with OH^- in the air to form nitric acid:

$$NO_2 + OH^- \rightarrow HNO_3$$

Nitric acid is a very strong acid, with a large $K_w = 10^1$, which results in very fast dissociation as soon as it is in contact with water in the atmosphere:

$$HNO_3 + H_2O \rightarrow H_3O^+ + NO_3^-$$

All living matters content phosphorus, but its concentration in many natural water environments is low. Phosphoric acid (H3PO4), which is not very volatile, can lose up to 3 H^+:

$$H_3PO_4 = H^+ + H_2PO_4^-$$

$$H_2PO_4^- = H^+ + HPO_4^{2-}$$

$$HPO_4^{2-} = H^+ + PO_4^{3-}$$

Sulfate occurs in natural water as organic sulfur, hydrogen sulfide (H_2S), elemental sulfur (S) and sulfate (SO_4^{2-}). Hydrogen sulfide (H_2S) is toxic for many organisms and is a source of odour in water. Hydrogen sulfide (H_2S) can also combine and precipitate heavy metals such as iron, zinc and cobalt. These metal elements are required for bacterial growth, therefore high levels of H2S may inhibit growth.

2.2.2 Water Treatment Processes

Preliminary Treatment

In this treatment, coarse solids and suspended impurities are removed by passing waste wafer through ball and mesh screens.

Primary Treatment or Settling Process

In this treatment, geyser proportion of suspended inorganic and organic solids is removed from liquid sewage by settling. In order to facilitate coagulants like alum, ferrous sulfate. These produce large gelatinous precipitate, which entrap timely divided organic matter and settle.

$$Al_2\left(SO_4\right)_2 + 6\,H_2O \rightarrow 2\,Al\left(OH\right)_3 \downarrow + 3\,H_2SO_4$$

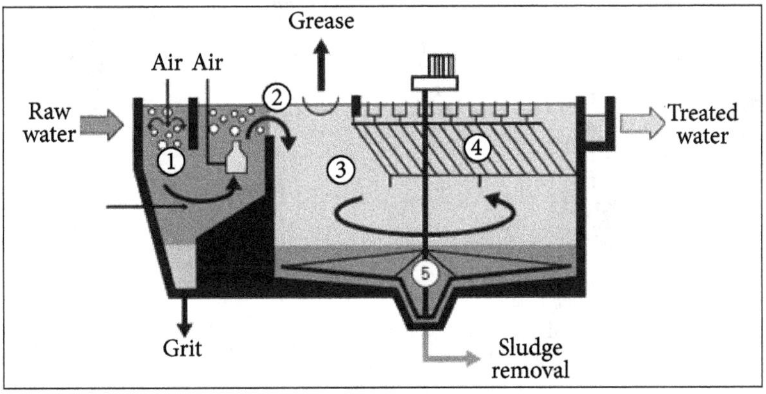

Primary treatment plant.

Secondary or Biological Treatment

It removes up to 90% of O2 demanding wastes. This is done by trickling filter or activated sludge process.

Trickling Filter Process: It is a circulated tank and is filled with either coarse on rock by means of slowly rotating arms.

When sewage starts percolating downwards grown on surface of filtering media using material of sewage as food. After completion of aerobic oxidation the sludge is removed. This process removes about 80-85% of BOD.

Activated Sludge Process: This is biologically active sewage and has large number of aerobic bacteria's which can easily oxidize the organic impurities. Then mixture is aerated in aeration tank. Under these conditions organic impurities of sewage get oxidized by micro-organisms. After aeration, sewage is taken by sedimentation tank. Activation sludge, portion of which used for seeding fresh batch of sewage. This removes 90-95% of BOD.

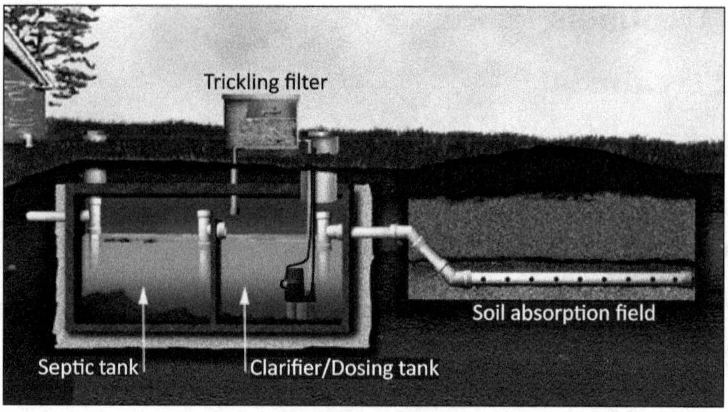

Trickling Filter

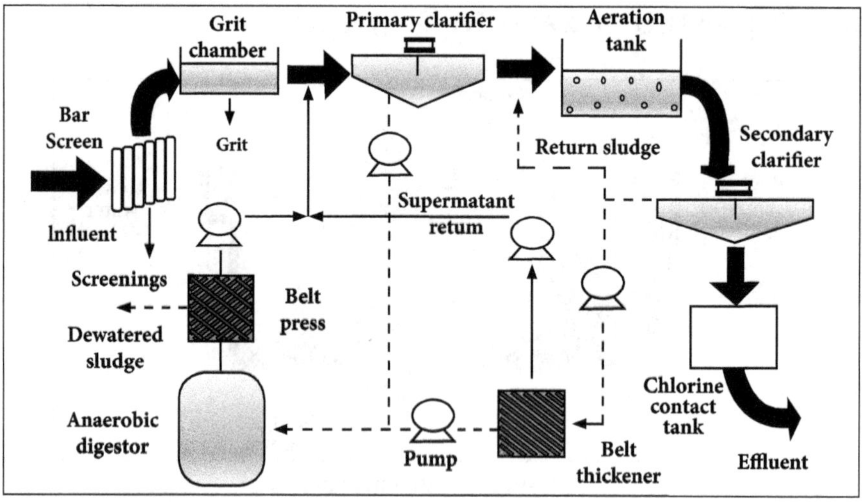

Activated Sludge Process

Tertiary Treatment: After secondary treatment, sewage effluent has a lower BOD (25 ppm), which can be removed by tertiary treatment process.

In tertiary treatment, effluent is introduced into flocculation tank, where lime is added to remove phosphates from the effluent led to ammonia stripping tower, pH maintained to 11 and the NH4 is converted together NH_3. The activated charcoal column, where wastes are adsorbed by charcoal finally, water is treatment with disinfectant (chlorine).

Disposal of Sludge

- Dumping into low-lying areas,
- Building of sludge (incineration),
- Dumping into sea,
- Using it as low grade fertilizers.

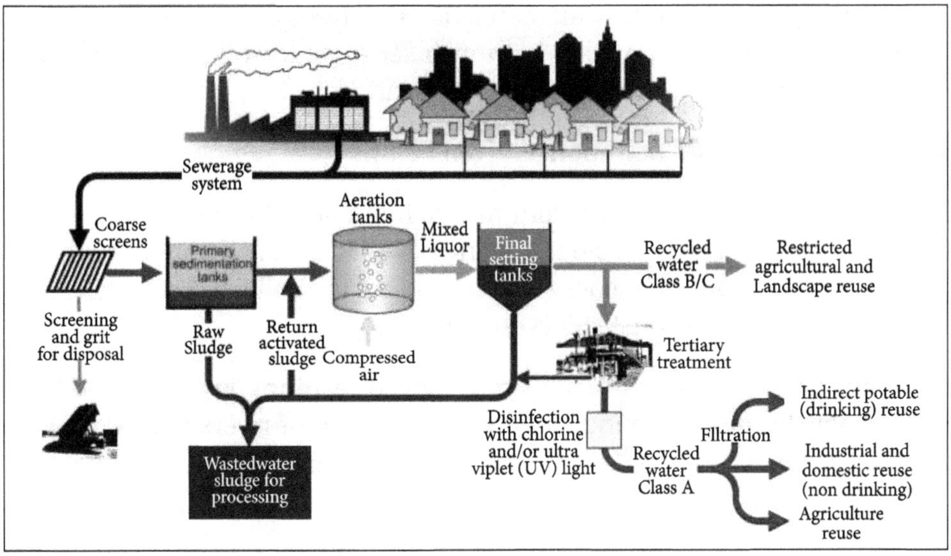

Sewage treatment plant.

2.3 Soil Pollution

Soil pollution is caused due to the addition of chemicals which reduce its productive capacity. In addition to the urban solid wastes, several hazardous chemicals are dumped into the soil. The toxic substances from these dumps leach out and percolate through the soil and contaminate the ground water. In agricultural operations, most of fertilizers and pesticides are used and the residual chemicals remain in the top layers of soil. Toxic insecticides kill useful soil bacteria that are favorable for plant growth.

Soil acts like a filter in removing the impurities in water and waste waters. But toxic residual chemicals from the soil reach human beings through vegetables, fruits, etc. Industrial effluents and solid wastes without adequate treatment which get deposited on the land or in water bodies are major sources of soil pollution.

Various pollutants and their harmful effects on soil are as follows:

- Organic wastes enter the soil pores and decompose. Pathogenic bacteria present in these wastes spread infection. Hookworm and also cause diseases.
- Compounds containing arsenic, mercury, chromium, nickel, lead, cadmium, zinc and iron are toxic to life. Fluorides also affect the plant growth.
- Excess use of sodium, magnesium, calcium, potassium, sulfur, zinc and iron in the form of fertilizers and pesticides inhibit plant growth and reduce crop yield. It is necessary to keep the dosage at an optimal level.

- Water logging and salinity increases the dissolved salt content in ground water and the soil. Some plants are too much sensitive to soil pH and salinity. High salinity may make the land unfit for cultivation.

Soil erosion also causes soil pollution as given below:

Soil erosion is the removal of soil from its original place. Water, ice wind and other climatic agents promote soil erosion. Cutting of trees, construction and other human activities reduce the vegetation cover which protects the soil from sun, rain, wind and moving water and as a result the soil is eroded.

Due to soil erosion essential minerals are removed resulting in the loss of fertility of soil. Soil erosion results in the increased sedimentation of rivers and lakes which has an impact on water quality and subsequently on the populations of aquatic organisms including fish.

It is not advisable to construct any new building on soils such as clay and silt that have been recently carried forward and deposited by the waters of rivers and streams as the load bearing capacity of such a soil medium is very low. There are cases where the underground stratum is peat deposited by sewage canals that had existed before. Such sites should never be selected for the construction of buildings.

In a natural ecosystem, essential nutrients and minerals cycle from the soil to living organisms and back to the soil when the organisms die. This natural s system is disrupted by agricultural operations. In dry climates much of the water used for irrigation evaporates, leaving behind high concentration of salts, such as sodium chloride, in the top soil, The accumulation of salts in the top soil is known as salinization.

For the conservation of soil, various methods are used to reduce soil erosion, to prevent depletion of soil nutrients and to restore nutrients already lost by erosion, leaching and excessive crop harvesting.

2.4 Marine Pollution

Marine pollution occurs when harmful or potentially harmful, effects result from the entry into the ocean of chemicals, particles, industrial, agricultural and residential waste, noise or the spread of invasive organisms. Most sources of marine pollution are land based. The pollution often comes from non-point sources such as wind-blown debris, agricultural runoff and dust. Nutrient pollution is a form of the water pollution that refers to contamination by excessive inputs of nutrients. It is a primary cause of eutrophication of surface waters, in which excess nutrients, usually phosphorus or nitrogen, stimulate algae growth.

Causes of Marine Pollution

Toxic Chemicals from Industries

Industrial and agricultural waste is another most common form of wastes that are directly discharged into the oceans, resulting in ocean pollution. The dumping of toxic liquids into the ocean directly affects the marine life as they are considered hazardous and secondly, they raise the temperature of the ocean, known as thermal pollution, as the temperature of these liquids is quite high. Animals and plants cannot be survived at higher temperatures eventually perish.

Sewage

Sewage or polluting substances flow through sewage, rivers or drainage's directly into the ocean. This is often how the minerals and substances from mining camps find their way into the ocean. The release of other chemical nutrients into the ocean's ecosystem leads to the reductions in oxygen levels, the decay of plant life, severe decline in the quality of sea water itself. As a result, all levels of oceanic life, plants and animals are highly affected.

Large Scale Oil Spills

Ship pollution is a huge source of ocean pollution, the most devastating effect of which is oil spills. Crude oil lasts for years in the sea and is extremely toxic to marine life, often suffocating marine animals to death once it entraps them. Crude oil is also extremely difficult to clean up, unfortunately meaning that when it is split.

Land Runoff

Land runoff is another source of pollution in the ocean. This occurs when the water infiltrates the soil to its maximum extent and the excess water from rain, flooding or melting flows over the land and into the ocean. Often times, this water picks up man-made, harmful contaminants that pollute the ocean, including petroleum, fertilizers, pesticides and other forms of the soil contaminants. Fertilizers and waste from the land animals and humans can be a huge detriment to the ocean by creating dead zones.

Littering

Pollution from the atmosphere is, believe it or not, a huge source of ocean pollution. This occurs when objects that are far inland are blown by the wind over long distances and ends up in the ocean. These can be anything from natural things like dust and sand, to man-made objects such as debris and trash. Most debris, especially plastic debris, cannot decompose and remains suspended in the oceans for years. Animals can become snagged on the plastic or mistake it for food, will slowly kill them over a long period of time. Animals who are most often the victims of plastic debris include turtles, dolphins, fish, sharks, crabs, sea birds and crocodiles.

In addition, the temperature of the ocean is highly affected by carbon dioxide and climate changes, which impacts the ecosystems and fish communities that live in the ocean. In particular, the rising levels of Co_2 acidify the ocean in the form of acid rain. Even though the ocean can absorb carbon dioxide that are originated from the atmosphere, the carbon dioxide levels are steadily increasing and the ocean's absorbing mechanisms, due to the rising of the ocean's temperatures, are unable to keep up with the pace.

Ocean Mining

Ocean mining in the deep sea is yet another source of ocean pollution. Ocean mining sites drilling for silver, gold, copper, cobalt and zinc create sulfide deposits up to three and a half thousand meters down in to the ocean. While we have the gathering of scientific evidence to explain the harsh environmental impacts of deep sea mining, we do have that deep sea mining causes damage to the lowest levels of ocean and increases the toxicity of the region. This permanent damage also causes leaking, corrosion and oil spills that only drastically further hinder the ecosystem of the region.

Effects of Ocean Pollution:

- Disruption to the Cycle of Coral Reefs.
- Effect of Toxic Wastes on Marine Animals.
- Failure in the Reproductive System of Sea Animals.
- Depletes Oxygen Content in Water.
- Affects Human Health.
- Effect on Food Chain.

2.5 Noise Pollution

The unwanted, unpleasant or disagreeable sound that causes discomfort for all living things.

The sound intensity is measured in decibel (dB), which is tenth part of the longest unit Decibel, One dB is equal to the faintest sound, a human ear can hear.

Noise pollution.

Types of Noise

It has been found that the environmental noise is being doubling for every 10 years. Generally noise is described as:

Industrial Noise

Highly intense sound or noise pollution is caused by many machines. There exists a long list of sources of noise pollution including different machines of numerous factories, industries and mills.

Industrial noise, particularly from mechanical saws and pneumatic dwell is unbearable and nuisance to public.

Example: In the start industry, the workers near the heavy industrial blowers are exposed to 112 dB for eight hours and suffer from the occupational pollution.

Transport Noise

The main noise comes from transport. It mainly includes road traffic noise, rail traffic noise and air craft noise. The number of road vehicles like motors, scooters, cars and particularly the diesel engine vehicles have increased enormously in recent years.

A survey conducted in metropolitan cities has shown that noise level on Delhi, Bombay and Calcutta is as high as 90 dB. Inhabitants of cities are subjected to the most annoying form of transport noise which gradually deafens them.

Neighborhood Noise

This type of noise includes disturbances from the household gadgets and community. Common noise makes are musical instruments, TV, VCR, radios, transistors, telephones and loudspeakers, etc. Ever since the industrial revolution, noise in environment has been doubling every ten years.

Effects of Noise Pollution

- It causes the muscles to contract leading to nervous breakdown, tension, etc.
- Noise pollution affects human comfort, health and efficiency. It causes contraction of blood vessels, makes the skin pale, leads to excessive secretion of the adrenalin hormone into blood stream which is responsible for high blood pressure.
- It affects health efficiency and behavior. It may cause damage to heart, brain, kidneys, lives and may also produce emotional disturbances.
- The adverse reactions are coupled with a change in hormone content of blood, which in turn increase the rate of heart beat, construction of blood vessels, digestive spasms and dilation of pupil of eye.
- Recently it has been reported that blood is also thickened by excessive noises.
- In addition to the serious loss of hearing due to excessive noise, impulsive noise also causes psychological and pathological disorders.

Control Measures

- Source Control: This may include source modification such as acoustic treatment to machine surface, design changes, limiting the operational timings and so on.
- Transmission Path Intervention: This may include containing the source inside a sound insulating enclosure, construction of noise barrier or provision of sound absorbing materials along the path.
- Receptor Control: This includes protection of the receives by altering the work schedule or provision of personal protection devices such as ear plugs for operating noisy machinery. The measure may include dissipation and deflection methods.
- Oiling: Proper oiling will reduce the noise from the machines.

Preventive Measures

Noise can be reduced by prescribing noise limits for vehicular traffic, ban on honking of horns in certain areas and creation of silent zones near schools and hospitals and redesigning of buildings to make them noise proof. Other measures can involve reduction of traffic density in residential areas and giving preferences to the mass public transport system.

2.6 Thermal Pollution

Addition of excess of undesirable heat to water that makes it harmful to man, animal or aquatic life or otherwise causes significant departures from the normal activities of aquatic communities in water.

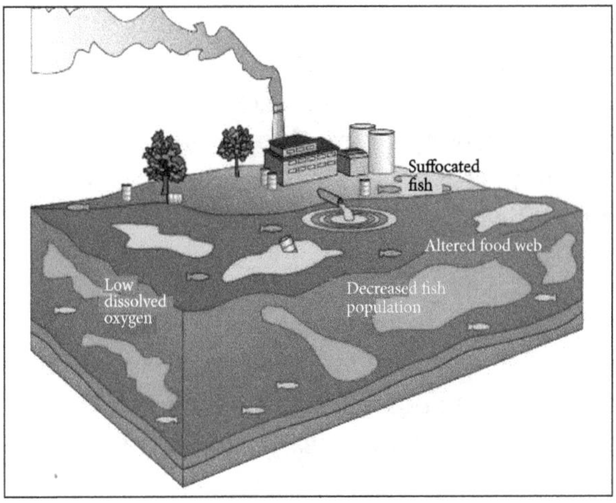

Thermal pollution.

Sources of Thermal Pollution

The following sources contribute to thermal pollution:

- Nuclear Power Plants: Nuclear power plants including drainage from hospitals, research institutes, nuclear experiments and explosions, discharged a lot of unutilized heat and traces of toxic radio nuclides into nearby water streams. Emissions from nuclear factors and processing installation are also responsible for increasing the temperature of the water bodies. The operations of power reactors and nuclear fuel processing of power reactors and nuclear fuel processing units constitute the major contributor of heat in the aquatic environment.

  Heated effluents from power plants are discharged at 10°C, higher than the receiving water which affects the aquatic flora and fauna.

- Coal-Fired Power Plants: Coal-fired power plants constitute the major source of thermal pollutants. Their condenses coils are cooled with water from nearby lake or river and discharge the hot water back to the stream increasing the temperature of nearby water to about 15°C. The heated effluents decrease the dissolved oxygen contents of water. It results into killing of fish and other marine organisms.

- Industrial Effluents: Industries generating electricity like coal powered and nuclear powered plants require huge amounts of cooling water for heat removal. Other industries like textile, paper and pulp as well as sugar also release heat in water but to a much lesser extent. Normally the discharged water from stream-electric power industry using turbo generators will have a higher temperature ranging from 6°C to 9°C than the receiving water. This results in the increase of stream temperature to a level at which natural dissipation of heat will be inefficient.

- Domestic Sewage: Domestic sewage is commonly discharged into rivers, lakes, canals or streams with or without waste treatment. The municipal sewage normally has a higher temperature than the receiving water. With the increase in temperature of the receiving water, the dissolved oxygen content decreases and demand of oxygen increases. The marine organisms which depend on the dissolved oxygen will die out.
- Hydroelectric Power: Generation of hydroelectric power. Sometimes, results in negative thermal loading in water systems. Apart from electric power industries, various factories with cooling requirement contribute to thermal loading. It has been reported that about 18% more heat is given to cooling waters in nuclear power plants than any other plant of equivalent size.

Effects of Thermal Pollution

- Reduction in dissolved oxygen: Concentration of dissolved oxygen decreases with increase in temperature of water.
- Increase in Toxicity: The rising temperature increase the toxicity of the poison present in water. A 10°C rise in temperature doubles the toxic effect of potassium cyanide. While a 80°C rise in temperature triples the toxic effects of oxygen causing massive mortality of fish.
- Food Shortage for Fish: Change in temperature alters the seasonal variation in the type and abundance of lower organisms. This fish may lack the right food at the right time.

2.7 Nuclear Hazards

Nuclear hazard is defined as the risk or danger to human health or the environment posed by radiation emanating from the atomic nuclei of a given substance or the possibility of an uncontrolled explosion originating from a fusion or fission reaction of atomic nuclear.

Sources of Nuclear Hazards

The sources of radioactivity are both natural and manmade.

The natural sources include:

- Cosmic rays from outer space: The quantity depends on altitude and latitude. It is more at higher latitudes and high altitudes.
- Emissions from radioactive materials from the Earth's crust: People have been

exposed to low levels of radiation from these natural sources for several. But it is the manmade sources which are posing a threat to mankind. The man-made sources of radioactivity are nuclear wastes produced at the time of:

- Uses of radioactive material in nuclear power plants.
- Mining and processing of radioactive ores.
- Uses of radioactive materials in nuclear weapons.
- Uses of radioactive isotopes in medical, industrial and research applications.

The greatest exposure to human beings comes from the diagnostic use of X-rays, radioactive isotopes used as the tracers and the treatment of cancer and other ailments.

Effects of Nuclear Hazards

The effects of radioactive pollutants depend upon half-life, energy releasing capacity, rate of diffusion and rate of deposition of the contaminant. Various atmospheric conditions and climatic conditions such as temperature, wind and rainfall also determine their effect. All organisms are affected from radiation pollution and the effects are extremely dangerous. The effects may be somatic or genetic damage. The effects are cancer, shortening of life span and genetic effects or mutations. Some of the possible effects are listed as below:

- Exposure at low doses of radiations (100-250 rads), men do not die but begin to suffer from fatigue, nausea, vomiting and loss of hair. But recovery can be possible.
- Radiations may break the chemical bonds such as DNA in cells. This affects the genetic makeup and control mechanisms. The effects can be instantaneous, prolonged or delayed types. Even it could be carried to future generations.
- Higher irradiation doses (10,000rads) kills the organisms by damaging the tissues of heart, brain, etc.
- Exposure at higher doses (400-500 rads), blood cells are reduced, bone marrow is affected, blood fails to clot and the irradiated person soon dies of infection and bleeding.
- Through food chain also, radioactivity effects are experienced by man. But the most significant effect of radioactivity causes long range effects, natural resistance and fighting capacity against germs is reduced, affecting the future of man and hence the future of our civilization.
- Workers handling the radioactive wastes get slow but continuous irradiation and in course of time develop cancer of different types.

Control of Nuclear Hazards

The peaceful uses of radioactive materials are so wide and effective that modern civilization cannot go without them on the other hand; there is no cure for radiation damage. Hence, the only option against nuclear hazards is to check and prevent radioactive pollution.

- Leakages from nuclear reactors, careless handling, transport and use of radioactive fuels, fission products and radioactive isotopes have to be totally stopped.
- Safety measures are enforced strictly.
- Waste disposal is careful, efficient and effective.
- There should be regular monitoring and quantitative analysis through frequent sampling in the risk areas.
- Preventive measures should be followed so that background radiation levels do not exceed the permissible limits.
- Appropriate steps should be taken against the occupational exposure.
- Safety measures should be strengthened against the nuclear accidents.

Disposal of Nuclear Wastes

Since nuclear waste can be extremely dangerous and therefore, the way in which they are to be disposed of is strictly controlled by international agreement. Since 1983, by international agreement, the disposal in the Atlantic Ocean and into the atmosphere has been banned. After processing, to recover the usable material and reducing the radioactivity of the waste, disposal is made in solid form where possible.

The nuclear wastes are classified into three categories such as:

- High Level Wastes (HLW): High level wastes have a very high radioactivity per unit volume. For example, spent nuclear fuel. have to be cooled and are, therefore, stored for several decades by its producer before disposal. Since these wastes are too dangerous to be released anywhere in the biosphere, they must be contained either by converting them into inert solids and then buried deep into earth or are stored in deep salt mines.
- Medium level wastes (MLW): Medium level wastes are solidified and are mixed with concrete in steel drums before being buried in deep mines or below the sea bed in the concrete chambers.
- Low liquid wastes (LLW): Low liquid wastes are disposed of in steel drums in concrete lined trenches in designated sites.

2.8 Role of an Individual in Prevention of Pollution

Environmental pollution cannot be prevented and removed. The proper implementation and especially the individual participation are the important aspects which should be given due importance.

The individual participation is useful in law making processes and restraining the pollution activities and thereby the public participation plays a major role in the effective environmental management. A small effect is made by each individual at his own place will have pronounced effect at the global level. It is aptly said “Think globally act locally.”

Each individual should change his or her life style in such a way as to reduce environmental pollution.

Individual Participation

- Help more in pollution prevention than pollution control.
- Plant more trees.
- Purchase recyclable, recycled and environmentally safe products.
- Use water, energy and other resources efficiently.
- Use natural gas than coal.
- Use CFC free refrigerators.
- Increase use of renewable resources.
- Reduce deforestation.
- Use office machines in well ventilated areas.
- Remove Nox from motor vehicular exhaust.

2.9 Field Study: Local Polluted Urban, Rural, Industrial and Agricultural Sites

A Visit to Local Polluted Sites

Today, the cry of pollution is heard from all the nooks and concerns of the globe and pollution has become a major threat to the very existence of mankind on this earth. It is the major challenge of the present time. The pollution of various resources has gone to such an extent that we are unable to breath fresh air and drink fresh water.

Today, different types of pollutants are piling up on the earth surface. We hardly find any place bee from pollutants. The pollutants differ both in quality and quantity at different places and time. A visit to selected sites helps us to study the nature and quantity of pollutants at those sites.

Urban Site

The air of an urban site has a large amount of gaseous and particulate pollutants released from automobiles. The land site of urban area has a number of non-degradable wastes like plastics, polythene, different types of packing materials and domestic waste. The water bodies have a number pollutants like soap and detergents, toxic chemicals, insecticides etc.

Rural Site

The air of a rural site is comparatively less polluted as compared to the urban and industrial sites. However, the air near the villages may contain smoke of burning dung cakes specially during early morning and evening hours. Open and wastelands may possess domestic wastes such as polythene plastics etc. The water bodies like ponds and pools may contain chemical fertilisers and pesticides.

Agricultural Site

The air at agricultural site is comparatively clean but it may contain suspended particles of dust and straw during harvesting and thrashing seasons of crops. The land may contain organic wastes in the form of animal excreta and crop residue. The ponds and pools nearby agricultural fields may be polluted with chemical fertilizers and pesticides.

Industrial Site

The air at an industrial site is highly polluted with smoke and different types of toxic gases. Many times the area gives a hazy look due to the presence of suspended particles in the air released from the chimneys of the industry. The land and water bodies surrounding the industry is polluted by number of affluents released by the industry. The observations relating to the visit of polluted sites may be presented in the form of following table:

S. No.	Polluted sites	Types of pollutants	Effects of pollutants on			
			Animals	Plants	Humans	Environment
1.	Rural site					
2.	Urban site					
3.	Industrial site					
4.	Agricultural site					

Chapter 3

Natural Resources

3.1 Forest Resources

Forest, an important resource to mankind, supports in a direct or indirect way the development of human economy and society with its thousands of material or non-material products. The forest is also the main object for the continental ecological system.

The successful management of forest inheritance plays a vital role in managing the global environment, conserving the other related resources and guaranteeing the sustainable development of economy and society.

Over-Exploitation

Due to overcrowded population materials supplied by the forest like Medicines, Shelter and Fuel are not sufficient to meet the people's demand. Thus the exploitation of forest materials is going on increasing day by day.

With growing civilization, the demand for raw materials like timber, pulp, minerals, fuel wood, etc., increases resulting in large scale logging, mining, road building and deforestations.

Reason for Over Exploration in India

It has been estimated that in India the minimum density of forests required to maintain good ecological balance is about 33, all area. But, at present it is only about 22%. So over exploitation forest materials occur.

Causes of Over-Exploitation

Over exploitation of forest wealth in the developing countries occurs in the following ways:

- Increasing industrial activities.
- Increasing agricultural production.
- Increasing in demand of wood resources.

3.1.1 Deforestation

Deforestation is the process of elimination of or removal of forest resources due to many natural or manmade activities. In general deforestation means destruction of forests.

Causes of Deforestation

Developmental Projects:

Developmental projects cause deforestation in two ways:

- Through submergence of forest area under water.
- Destruction of forest area.
    - Examples: Big dams, hydroelectric projects, road construction, etc.

Hence, there is a need to discourage the undertaking of any development works in the forest area.

Mining Operations:

- Mining have a serious impact on forest areas. Mining operation reduces the forest area.
    - Example: Mica, Coal, Limestone, etc.

Raw Materials for Industries:

- Wood is the most important raw material for so many purposes.
- Example: For making boxes, ply woods, etc.

Fuel Requirements:

- In India both tribal and rural population is dependent on the forest for meeting their daily needs of fuel wood, which leads to the pressures on forest, ultimately to deforestation.

Shifting Cultivation:

- The replacement of natural forest ecosystem for mono specific tree plantation can lead to disappearance of number of plant and animal species.

Forest Fires:

- Forest fire is one of the major cause for deforestation. Due to human interruption and rise in ambient temperature.

Effects of Deforestation on the Investment

Since many people are dependent on the forest resources, deforestation will have the following social, ecological and economic effects.

- Soil Erosion: Deforestation also causes soil erosion, drought, landslides, floods. Natural vegetation acts as a natural barrier to reduce the wind velocity, this will in turn reduce soil erosion.
- Loss of Biodiversity:
    - Most of the species are very sensitive to any disturbance and changes.
    - When the plants no longer exist, animals depend on them for food and habitat become extinct.
- Loss of Food Grains:
    - As a result of soil erosion, the countries loose the food grains.

3.1.2 Case Studies

Jhum Cultivation

Jhum Agriculture or shifting agriculture has destroyed large number of hectare of forest tracts in North-Eastern states and Orissa. Jhum agriculture is subsidence agriculture in which tract of forest land is cleared by cutting trees and it is used for cultivation. After few years, when the productivity of the land decreases, cultivators abandon the land and clear the next tract. As a result of this practice, combined with the increasing population, there is rapid deforestation as more and more cultivator's clear forest to cultivate the land. Also, with increase in population, cultivators are forced to return to the previous tracts of land in relatively shorter durations, not allowing the land to regain its productivity.

Chipko Movement

The Chipko movement or Chipko Andolan is a social-ecological movement that practised the Gandhian methods of Satyagraha and non-violent resistance, through the act of hugging trees to protect them from being felled. The modern Chipko movement started in the early 1970s in the Garhwal Himalayas of Uttarakhand, with growing awareness towards rapid deforestation. The landmark event in this struggle took place on March 26, 1974, when a group of peasant women in Reni village, Hemwal ghati, in Chamoli district, Uttarakhand, India, acted to prevent the cutting of trees and reclaim their traditional forest rights that were threatened by the contractor system of the state Forest Department. Their actions inspired hundreds of such actions at the grassroots level throughout the region. By the 1980s the movement had spread throughout India and led to formulation of people-sensitive forest policies, which put a stop to the open felling of trees in regions as far reaching as Vindhyas and the Western Ghats.

Western Himalayan Region

Over the last decade, there has been widespread destruction and degradation of forest resources in Himalayas, especially western Himalayas. This has resulted in various problems such as erosion of top soil, irregular rainfall, changing weather patterns and floods. Construction of roads on hilly slopes, have not only undermined their stability, but also damaged protective vegetation and forest cover. Tribes in these areas are increasingly facing shortage of firewood and timber, due large scale tree cutting. Increased traffic volumes on these roads leads to increased pollution in the area.

3.1.3 Timber Extraction

There has been unlimited exploitation of timber for commercial use. Due to increased industrial demand, timber extraction has significant effect on forest and tribal people.

Logging

- Poor logging results in degraded forest and may leads to soil erosion especially on slopes.
- There is a loss of long term forest productivity.
- New logging roads permit shifting cultivators and fuel wood gatherers to gain access to the logging area.
- Exploitation of tribal people by contractor.
- Species of plants and animals may be eliminated.

3.1.4 Mining

Mining from shallow deposits is done by surface mining while that from deep deposits is done by sub-surface mining. It leads to the degradation of lands and loss of the top soil. It is estimated that about eighty thousands hectare land is under stress of mining activities in India.

Mining leads to drying up perennial sources of water sources like streams and spring in mountainous area. Mining and other associated activities remove vegetation along with underlying soil mantle, which results in the destruction of topography and landscape in the area. Large scale deforestation has been reported in Mussorie and Dehradun valley due to the indiscriminating mining.

The forested area has declined at an average rate of about 33% and the increase in non-forest area due to mining activities has resulted in relatively unstable zones leading to landslides. Indiscriminate mining in forests of Goa since 1961 has destroyed more than 50000ha of forest land. The coal mining in Jharia, Raniganj and Singrauli areas has caused extensive deforestation in Jharkhand.

Mining of magnetite and soapstone have destroyed 14ha of forest in hilly slopes of Khirakot, Kosi valley and Almora.

Mining of the radioactive minerals in Kerala, Tamilnadu and Karnataka are posing similar threats of deforestation.

The rich forests of Western Ghats are also facing the same threat due to mining projects for excavation of copper, chromites, bauxite and magnetite.

3.1.5 Dams and their Effects on Forests and Tribal People

Pandit Jawaharlal Nehru referred dam and valley projects as "Temples of modern India". These big dams and rivers valley projects have multi-purpose uses. However, these dams are also responsible for the destruction of forests. They are responsible for loss of flora and fauna, degradation of catchment areas, disturbance in forest ecosystems and increase of water borne diseases, rehabilitation and resettlement of tribal peoples.

India has more than 1550 large dams, the maximum being in the state of Maharashtra, followed by Gujarat and Madhya Pradesh.

The highest one is Tehri dam, on river Bhagirathi in Uttaranchal and the largest in terms of capacity is Bhakra dam on river Satluj in Himachal Pradesh. Big dams have been in sharp focus of various environmental groups all over the world, which is mainly because of several ecological problems including deforestation and socioeconomic problems related to tribal or native people associated with them.

The Silent valley hydroelectric project was one of the first such projects situated in the tropical rain forest area of Western Ghats which attracted much concern of the people.

The crusade against the ecological damage and deforestation caused due to Tehri dam was led by Shri. Sunder Lal Bahaguna, who was the leader of Chipko Movement.

The cause of Sardar Sarovar Dam related issues have been taken up by the environmental activitist Medha Patkar, joined by Arundhati Ray and Baba Amte. For building big dams, large scale devastation of forests takes place which breaks the natural ecological balance of the region.

Droughts, floods and landslides become more prevalent in such areas. Forests are the repositories of invaluable gifts of nature in the form of biodiversity and by destroying them; we are going to lose these species even before knowing them.

These species could be having marvelous economic or medicinal value and deforestation results in loss of this storehouse of species which have evolved over millions of years in a single stroke.

Effects of Dams

Dams are massive artificial structures built across the river to create a reservoir in order to store water for many beneficial purpose. However, these dams are also responsible for the destruction of forest and displacement of local people.

The Indian Scenario

India has more than 1600 large dams. Dam is the highest built across the river in the State.

State	No. of Dams
Maharashtra	More than 500 dams
Gujarat	More than 250 dams
Pradesh	More than 130 dams

Effects of Dam on Forest

- Thousands of hectares of forest have been cleared for executing river valley projects.
- In addition to the dam construction, the forest is also cleared for residential accommodation, office buildings, etc.
- The big river valley projects also cause water logging which leads to salinity and in turn reduces the fertility of the land.
- Hydroelectric projects also have led to widespread loss of forest in recent years.
- Hydroelectric projects provide opportunities for the spread of water borne diseases.
- Construction of dams under these projects leads to killing of wild animals and destroying aquatic life.

Examples:

- Narmada Sugar Project: It has to submerged 3.5 each hectares of forest comprising of teak and bamboo trees.
- Dam: It has submerged 1000 hectares of forest affecting about 430 species of plants.

Effects of Dam on Tribal People

- The greatest social cost of big dam is the widespread displacement of tribal people such that biodiversity cannot be tolerated.
- The displacement and cultural change affects the tribal people both mentally and physically. They do not accommodate the modern food habits and life styles.
- Tribal people and their culture cannot be questioned and destroyed.
- Generally, the body conditions of the tribal people will not suit with the new areas and hence they will be affected by many diseases.

Conflicts of Water Resources

Conflicts through use:

- Unequal distribution of water led to interstate or international disputes.
- International conflicts: India & Pakistan fight to get water from the Indus.

Construction of Dams/Power stations:

- For hydroelectric power generation, dams built across the rivers, initiates conflict between the states.

Conflict through pollution:

- Rivers & Lakes are used for electricity, shipping & for industrial purpose.
- Disposal of waste water & industrial waste decrease the quality of water & causes pollution.

3.2 Water Resources

The water cycle, through evaporation and precipitation, maintains the hydrological systems which form rivers and lakes and supports in a variety of aquatic ecosystems. Wetlands are intermediate forms between the terrestrial and aquatic ecosystems and contain species of plants and animals that are highly moisture dependent. The aquatic ecosystems are used by several people for their daily needs such as drinking water, cooking, washing, watering animals and irrigating fields.

The world depends on a limited quantity of fresh water. Water covers 70% of the surface of earth but only 3% of this is fresh water. Of this, 2% is in polar ice caps and the remaining 1% is usable water in lakes, rivers and subsoil aquifers. Only a fraction of this is actually used. At a global level, 70% of water is used for agriculture, about 25% for industry and only 5% for domestic use. However this varies in different countries and industrialized countries use a greater percentage for industry. India uses 90% for agriculture, 7% for industry and 3% for domestic use. One of the greatest challenges facing the world in this century is the need to rethink overall management of water resources.

Floods

Floods have been a serious environmental hazard for a centuries. However, the havoc raised by rivers overflowing their banks has become progressively more damaging, as people have deforested catchments and intensified use of river flood plains that once acted as safety valves. Wetlands in the flood plains are nature's flood control systems into which the overfilled rivers could spill and act as a temporary sponge holding water and preventing fast flowing water from damaging the surrounding land. Deforestation in the Himalayas causes floods that year after year kill people, damage crops and destroy homes in the Ganges and its tributaries and the Brahmaputra.

Rivers can change their course during floods and tons of valuable soil is lost to the sea. As the forests are degraded, rainwater not a longer percolates that slowly into the subsoil but, runs off down in the mountainside bearing large amounts of topsoil. This blocks the rivers temporarily but gives way as the pressure mounts allowing enormous quantities of water to wash down suddenly into the plains below. There, rivers swell, burst their banks and flood waters spread to engulf peoples' farms and homes.

Drought

In case of most arid regions of the world, rains are unpredictable. This leads to periods when there is a serious scarcity of water to drink, it use in farms or provide for an urban and industrial use. Drought prone areas are faced with irregular periods of famine. Agriculturists have no income in these years and as they have no steady income, they have a constant fear of droughts. India has 'Drought Prone Areas Development Programs', which are used in such areas to buffer the effects of droughts. Under these schemes, people are given wages in bad years to build the roads, minor irrigation works and plantation programs.

It is an unpredictable climatic condition and occurs due to the failure of one or more monsoons. It varies in frequency in different parts of our country. While it is not feasible to prevent the failure of the monsoon, good environmental management can reduce the ill effects. The scarcity of water during the drought affects homes, agriculture and industry. It also leads to the food shortages and malnutrition that especially affects children. Several measures can be taken to minimize the serious impacts of a drought.

However this should be done as a preventive measure so that if the monsoons fail, its an impact on the lives of the local people is minimized. In years when the monsoon is adequate, we use good supply of water without trying to conserve it and use the water judiciously. Hence, during a year when the rains are poor, there is no water even for drinking in the drought area. One of the factors that worsens the effect of drought is deforestation.

Once the hill slopes are denuded of forest cover rainwater rushes down the rivers and is lost. Forest cover permits the water to be held in the area allowing it to seep into the ground. This charges the underground stores of water in natural aquifers. This can be used in drought years if the stores have been filled during a good monsoon. If water from an underground stores is overused the water table drops and vegetation suffers. This soil and water management and afforestation are the long term measures that reduces the impact of droughts.

Sustainable Water Management

'Save water' campaigns are essential to make people everywhere aware of dangers of water scarcity. A number of measures are to be taken for the better management of the world's water resources. These include measures such as:

- Building several small reservoirs instead of few mega projects.
- Soil management, micro catchment development and afforestation permits recharging of underground aquifers thus reducing the need for large dams.
- Preventing loss in Municipal pipes.
- Treating and recycling municipal waste water for an agricultural use.
- Preventing leakages from dams and canals.
- Develop small catchment dams and protect wetlands.
- Effective rain water harvesting in urban environments.
- Pricing water at its real value makes people use it more responsibly and efficiently and reduces water wasting.
- Water conservation measures in agriculture namely using drip irrigation.
- In the deforested areas where land has been degraded, the soil management by bunding along the hill slopes and making 'nala' plugs can help to retain moisture and make it possible to revegetate degraded areas. Managing a river system is best done by leaving its course as undisturbed as possible. Dams and canals may leads to major floods in the monsoon and the drainage of wetlands seriously affects the areas that get flooded when there is high rainfall.

3.2.1 Use and Over Utilization of Surface and Ground Water

Main sources of water are:

- Surface water sources: Lakes impounding reservoirs, streams, seas, irrigation canals.
- Ground water sources: Springs, wells, infiltration wells.
- Merits of ground water sources:
    - The water quality is good and better than surface source.
    - Being underground, the ground water supply has less chance of being contaminated by atmospheric pollution.
    - The land above ground water source can be used for other purposes and has less environmental impacts.
    - Prevention of water through evaporation is ensured and thus loss of water is reduced.
    - Ground water supply is available and can even be maintained in deserted areas.

Demerits of ground water source:

- Modeling, analysis and calculation of ground water is less reliable and based on the past experience, thus posing high risk of uncertainty.
- Water obtained from ground water source is always pressure less. A pump is required to take the water out and is then again pumped for daily use.
- The transport / transmission of ground water is a problem and an expensive work. Water has to be surfaced or underground conduits are required.
- Boring and excavation for finding and using ground water is expensive work.

Impounding Reservoirs

It is a basin constructed in the valley of a stream or river for the purpose of holding stream flow so that the stored water may be used when supply is insufficient.

They have the following two functions:

- To retard the flood.
- To impound water for beneficial use.

Two functions may be combined to some extent by careful operations.

An impounding reservoir presents a water surface for evaporation. This loss must be considered. Possibility of large seepage loss must also be considered. If it is economically impossible to prevent them, the project may have to be abandoned or move it to a more favorable site. There will be some loss by seepage through and under the dam itself.

Well Hydraulics

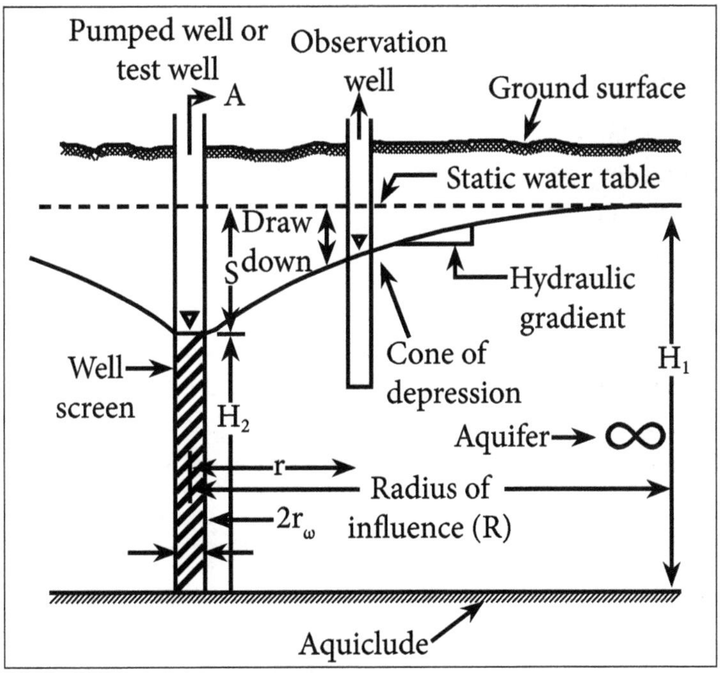

Well Hydraulics.

Fresh water is a natural and renewable resource. Rate at which it is renewed is becoming surpassed by the rate that is being consumed in the Tampa Bay area. Job of finding new sources of fresh water and managing the existing sources have fall under authority of the Southwest Florida Water Management District.

Agriculture Needs of Water for the Crops

As our country is an agricultural based country, the crops are to be developed for the production of different types of foods grains. The requirement of water varies from crop to crop. Different research stations are busy in identifying the water needs of all the crops. The most of these crops are shallow rooted, thus water being extracted from to top layers of the soil. The soil moisture available in the top layers is essential for such crops. Hence, it is advocated to keep the topsoil always moist so that crops do not wilt under no water of condition.

Many scientific methods of ploughing are developed to maintain the moisture in the top-soil for longer periods. The requirement of maintaining the water storage for breeding the fish and prawn culture is attaining importance whenever we go for storage project. If since fish cannot travel against flow when the velocity is high, special structures are provided adjacent to the weirs to maintain low velocities. Such structures are known as fish ladders.

The different purposes of the projects have their conflicting affect on the environment. However, depending on the priority of the water needs of the area to be served, planning has to proceed in a way such that the impact on environment is kept to be minimum while serving the urgent and primary needs of the people of that area.

Groundwater and Aquifers

The water that is available in the deeper layers of the earth is known as Groundwater. This water has been trapped inside the earth's crust for several centuries. The water that is lying under the ground has the capacity to move in general direction of its slope with a very small velocity.

The bodies that contain such water are known as Aquifers. If water from these aquifers can be drawn by digging wells and pumping water from these wells. Some of these wells can supply very large quantities of water. In some cases, the drinking water needs of cities are completely met by the groundwater bodies. In the agriculture sector also, groundwater is supporting the growth of crops enormously in the agricultural dominated areas.

Groundwater and Wells

The wells are used to bring groundwater to the land surface by means of pumps. If wells can be deep wells and shallow wells depending upon the depth at which ground water is available. If sometimes open dug wells are used where the water table is high.

In the case of deeper and hard rock aquifers, Tube wells are constructed. In such cases, deep well pumps such as jet pumps or turbine pumps are used to lift water to the surface of the ground. These tube wells in many cases yield significantly high discharges.

In the way, groundwater is the water that is available at the place of its use. Hence, this can be lifted from the wells in the agricultural fields and supplied to the crops raised in that field. Hence, there is no need to send to water long distances. This incidentally reduces the water losses considerably. The use of ground water for nearby areas also does not pose the problems of environmental degradation.

3.2.2 Dams: Benefits and Problems

Dams

Today there are more than 45,000 large dams around the world, which provide an important role in the communities and economies that harness these water resources for their economic development. Current estimates suggest some 30-40% of irrigated land worldwide relies on the dams. Hydropower currently supplies 19% of the world's total electric power supply and is used in over 150 countries. The world's two most populous countries such as China and India have built around 57% of the world's large dams.

Dams Problems

- Serious impacts on riverine ecosystems.
- Fragmentation and physical transformation of rivers.
- Dislodging animal populations, damaging their habitat and cutting off their migration routes.
- Water logging and salinization of surrounding lands.
- Fishing and travel by boat disrupted.
- Social consequences of large dams is due to the displacement of people.
- The emission of greenhouse gases from reservoirs due to rotting vegetation and carbon inflows from the catchment is a recently identified impact.

Large dams have had serious impacts on the livelihoods, lives, cultures and spiritual existence of indigenous and tribal peoples. They have suffered disproportionately from the negative impacts of the dams and often been excluded from sharing the benefits. In India, of 16 to 18 million people displaced by dams, 40 to 50% were tribal people, who account for only 8% of our nation's one billion people. Conflicts over dams have heightened in the last two decades because of their social and environmental impacts and failure to achieve the targets for sticking to their costs as well as achieving promised benefits. Recent examples show how failure to provide a transparent process that includes effective participation of the local people has prevented affected people from playing an active role in debating the pros and cons of the project and its alternatives. The loss of traditional, local controls over equitable distribution remains a major source of conflict.

Benefits of Dams

- Hydroelectric generation.
- Transfer of water using canals from areas of excess to areas of deficit water.
- Irrigation during dry period.
- Flood control and soil protection.

- Ensuring year-round water supply.
- Multipurpose river valley projects also provide for inland water navigation.

Environmental Problems Created by Dams

- Submergence of large areas of land that might include fertile fields and human settlements.
- The enormous weight of water behind the dam could trigger seismic activity that might crack the dam and unleash a flood.
- A number of water-related diseases occur (e.g. malaria) particularly where water impoundments provide breeding sites for the vectors. The spread of onchocerciasis in populations living near dam spillways and downstream has also been seen.
- Resettlement and rehabilitation problem of the displaced people.
- Loss of free flowing rivers that are either drowned by reservoir impoundments or turned into linear, sterile irrigation canals.
- Growth of snail populations in the shallow permanent canals that distribute water to fields and human settlements.
- Salts left behind by evaporation increase the salinity of the river and make its water unusable when it reaches the downstream cities.
- Accumulating the sediments in the storage reservoir not only makes dams useless but also represents a loss of valuable nutrients to the downstream agricultural lands.

3.3 Mineral Resources: Use and Exploitation

Minerals are naturally occurring substances having definite chemical composition and physical properties.

Formation of Mineral Deposits

Concentration of the mineral at a particular spot, which can be extracted profitably and gives rise to a mineral deposit. The formation of these deposits is a very slow biological process, it even takes millions of years to develop as a mineral deposit.

Various Biological Processes

- Formation of mineral deposits is due to the biological decomposition of dead animals and organic matters.

- Mineral deposits also formed due to evaporation of sea water.
- Mineral deposits are formed due to oxidation reduction reaction inside the earth.

Classification of Mineral Resources

U.S. Geological Survey divides non -renewable mineral resources into 3 categories:

Identified Resources: The location, existence, quantity and quality of these mineral resources are known by the direct geological evidence and measurements.

Undiscovered Resources: These mineral resources are assumed to be exist on the basis of geological knowledge and they but their locations, quantity and quality are unknown.

Reserves: These mineral resources are identified resources, from which a usable minerals can be extracted profitably.

Uses of Minerals

- Development of industrial plants and machinery.
- Construction, housing, settlements.
- Generation of energy.
- Communication purposes.
- Medicinal purposes, particularly in Ayurveda system.

Classification of Minerals

Minerals are classified into two ways based on this composition and usage.

1. Based on Composition

Based on composition, minerals can be classified into two types.

- Metallic Minerals

  Metallic minerals are the one from which various kinds of metals can be extracted.

  Example: Iron, copper, zinc, etc.

- Non-Metallic Minerals

Non-metallic minerals are the one from which various non-metallic compound can be extracted.

Example: Quartz, folds par, calcite.

2. Based on Usage

Based on usage, minerals are classified into two types:

- Critical Minerals: These are essential for the economic power of a country. Example: Iron, Al, Cu and Au.
- Strategic Minerals: These are required for the defense of a country. Example: Manganese, cobalt.

Management of Mineral Resources

- The efficient used and protection of mineral resource.
- Modernization of the mining industries.
- Search for new deposit.
- Reuse and Recycling of the metals.
- Environmental impacts can be minimized by adopting the ecofriendly mining technology.

3.3.1 Environmental Effects of Extracting and Using Mineral Resources

Most important environmental concern arises from the impacts of extraction and processing of the minerals during mining, melting, roasting, etc.

Mining

Mining is the process of extraction of metals from a mineral deposit.

Types of Mining

The various types of mining are as follows:

- Surface mining: It is the process of extraction of raw materials from the near surface deposits.
- Underground mining: The process of extraction of raw materials below the earth's surface is known as underground mining. It includes,

- Open-Pit mining: In this type of mining machines dig holes and remove the ores. Example: Iron, Copper, Marble, etc.
- Dredging: In dredging, chained buckets and drag lines are used, which scrap up the minerals from wider-water mineral deposit.
- Strip Mining: In case of strip mining, the ore is stripped off by using bull-dozers, stripping wheels.

Environmental Damage

The environmental damage, caused by mining activities are as follows:

- DE vegetation and defacing of landscape: The top soil and the vegetation are removed from the mining area. Large scale deforestation or de vegetation leads to a number of ecological losses and also landscape gets badly affected.
- Ground water contamination: Mining disturbs and also pollutes the ground water. Usually sulfur, present as an impurity in many ores gets convert into disulphuric acid due to microbial action, which makes the water acidic. Some heavy metals also get leached into ground water.
- Surface water pollution: The drainage of acid from the mine often contaminates the nearby streams and lakes. The acidic water is harmful to many aquatic life. Radio active substances like Uranium also contaminate the surface water and kill many aquatic animals.
- Air pollution: Melting and roasting are done to purify the metals, which emits enormous amounts of air pollutants damaging the nearby vegetation. The suspended particulate matter, arsenic particles, cadmium, etc., contaminate the atmosphere and public suffer from several health problems.
- Subsidence of land: It is mainly associated with underground mining. Subsidence of mining area results in cracks in houses, tilting of buildings, bending of railway tracks.

Environment Damages

- Mining activity not only destroys trees, it also pollutes soil, water and air with heavy metal toxins that are almost impossible to remove.
- Due to the continuous removal of minerals, forest covers, the trenches are formed on ground, leading to water logged area, which in turn contaminates the ground water.
- Destruction of natural habitat at the mine and waste disposal sites.

- During mining operations, the vibrations are developed, which leads to earthquakes.
- Noise pollution is the another major problem from mining operations.
- When materials are disturbed in significant quantities during mining process, large quantities of sediments are transported by water erosion.
- Sometimes landslides may also occur as a result of continuous mining in forest area.
- Mining reduces the shape and size of the forest areas.

3.3.2 Case Studies

Gold Exploration in Ghana

Gold plays an important role in economic and cultural life of people. The colonial name of Ghana, 'Gold Coast' was based on the fact that when visitors from outside Africa first arrived on the shores of the Gulf of Guinea, they could literally see and pick gold from the sand. Today it is not impossible to do so, although it may require more effort.

Large-scale surface mining in particular involves the clearing of large tracts of forest and agricultural land, resulting in serious land and for. Degradation For the first time in the history of this country, some concerned chiefs in the Western Region went on a demonstration calling for a ban against gold mining in their part of the Region.

As recently as August 18, 1997 at a meeting with the Ghana Mine Workers Union, the Minister of Mines and Energy used the platform to appeal to Ghanaians to be more accommodating and to stop the call for the ban of surface mining. He admitted that there were some negative effects of mining operations, but he stressed that his Ministry would ensure that the environmental degradation characterizing such mining activities would be reduced to a minimum. The report did not indicate how this was going to be done. He also said that he would ensure that adequate compensation would be paid to the affected communities.

Our history is replete with reports of chiefs or Local Councils receiving these compensations and spending very little on amenities for the affected community. There is serious water pollution in the mining areas. Rivers and other water bodies that serve as the source of drinking water for the communities in the mining area or downstream from it complain of serious water pollution because of the mining activities. Recently then were reports of 'galamsey' operators having dug very dose to railway lines somewhere in the Eastern Region, thus putting the entire operation of the Accra-Takoradi railway services into jeopardy.

Galamsey operators also nearly destroyed a forest and game reserve in the Upper East Region. In the last two cases it took the intervention of the government, the police and the army to halt the operations. Air pollution is another serious side effect. Gold mines that are located underground usually harvest the vein type deposits. Gold in these deposits is associated with sulphur and arsenic and to some extent nitrogen compounds. These elements consume cyanide, the chemical that separates gold from its impurities. They must therefore be expelled before the cyanidation and this is done by roasting the powdered ore.

Heating converts the arsenic, nitrogen and sulphur into gaseous oxides that are contained in the emissions horn the stacks of the fumaces. Studies carried out in some of the mining areas and their surrounding communities reveal that the ambient air, the soil, riverbeds, fish, fresh vegetables and food items are heavily polluted with arsenic and mercury. Although since 1992, the Ashanti Goldfields Company Limited company has installed filters in its stacks, it is estimated that only 90 % of the harmful arsenic oxide is removed.

3.4 Food Resources

Food is an essential requirement for the human survival. Each person has minimum food requirement. The main components of food are carbohydrates, fats, proteins, minerals and vitamins.

Types of Food Supply

Historically humans have a dependency on three systems for their food supply:

- Croplands: It mostly produces the grains and provides around 76% of the world's food. Example: Rice, Wheat, Maize.
- Range lands: It produces food mainly from grazing livestock and provides around 17% of the world's food. Example: Meat, Milk, Fruits.
- Oceans: Oceanic fisheries supply about 7% of the world's food. Example: Fish, Prawn, Crab, etc.

3.4.1 World Food Problems

We know that 79% of the total area of the earth is covered with water. Only 21% of the earth surface is land, of which most of the areas are desert, forest, mountains, barren areas. Only a minimal percentage of land is for cultivation, which in turn often does not suffice for the ever growing world population, as day by day the world population increases, while the cultivable land area decreases.

The environmental degradation like soil erosion, water logging, water pollution, salinity, affect agricultural lands. Urbanization is another problem in developing countries, which deteriorates the agricultural lands.

Since the food grains like wheat, rice, corn and vegetable like potato are the major food for the people all over the world, the food problem raises.

A key problem is the human activity which degrade most of the earth is net primary productivity which supports all life.

We know that 79% of total area of the earth is covered with water. Only 21% of the earth surface is land, of which most of the areas are desert, forest, mountains, barren areas. Only a minimal percentage of land is for cultivation, which in turn often does not suffice for the ever growing world population, as day by day the world population increases, while the cultivable land area decreases.

Environmental degradation like soil erosion, water logging, water pollution, salinity, affect agricultural lands.

Urbanization is another problem in developing countries, which deteriorates the agricultural lands.

Since the food grains like rice, wheat, corn and the vegetable like potato are the major food for the people all over the world, the food problem raises.

A key problem is the human activity which degrades most of the earth is net primary productivity which supports all life.

Under Nutrition and Malnutrition

Nutrition or Nutritious or Nourished: To maintain good health and resist disease, we need large amount of the macro nutrients such as carbohydrates, proteins and fats and smaller amount of macro nutrients such as Vitamin A, C and E and minerals such as calcium, iron and iodine.

The food and agriculture organization (FAO) of United Nation estimated that on an average, the minimum caloric intake on a global scale is 2,500 calories/day.

Under Nutrition or Under Nourished: People who cannot afford enough food to meet the basic energy needs suffer from under nutrition. They receive less than 90% of the minimum dietary calories.

Effect of under nutrition: Suffer from mental retardation and infectious diseases such as and diarrhea.

Malnutrition or Malnourished: Besides the minimum caloric intake we also need proteins, minerals, vitamins, iron and iodide. Deficiency of lack of nutrition often lead to malnutrition resulting in several diseases.

Effect of Malnutrition

S. No.	Deficiency of Nutrient	Effects
1.	Proteins	Growth
2.	Iron	Anemia
3.	Iodide	Goiter, Cretinism
4.	Vitamin A	Blindness

Thus, chronically under nourished and malnourished people is disease prone and are too weak to work or think clearly.

World Food Problems

- Environmental degradation like soil erosion, water pollution, water logging, salinity, affect agricultural lands.
- Urbanization is another problem in developing countries, which deteriorates agricultural lands.
- A key protein is the human activity, which degrade most of the world's net primary productivity which supports all life.

3.4.2 Changes Caused by Agriculture and Overgrazing

Non-Agriculture Activities

The different types of non-farm activities can be performed by farmers in rural areas. These activities and their dominance vary from place to place. In the study sites, the predominant non-farm activities are trading and handicraft.

Trade is predominant in Kachabira Wereda, while handicraft, mainly weaving, is the predominant activity in Damotgale Wereda. Among the sample population, there are only six households who reported non-farm activity other than trading and handicraft in the study sites. There are three households reported sale of `Injera' and bread. Two farmers are employed as labourers while one farmer earns income from renting cattle.

Prior to 1974, population in Kambata and Welaita were known for their seasonal migration to plantation sites and commercial farms. The sample population did not indicate any of this pattern at present.

In a similar way, a study by Getachew (1995) also found the insignificance of labour migration at present. About 162 farmers or 70.1% reported that they earn income from non-farm activities. The remaining 67 farmers or 29.2% of our sample have no income from non-farm activity the mean income from non-farm activities is 633.95 birr.

The highest income is recorded for Lesho and Gemesha in Kachabira Weredas, where 1255.33 and 578.37 birr, respectively, were reported. Distribution of income from non-farm activities shows that about 46.2% earn in the range of 1-500 birr.

The drops to 16.2% for income group between 500 - 1000 birr. Those farmers earning above 1500 birr are only 5.7% of the total.

Changes Caused by Non-Agriculture Activities

An engagement in nonagricultural activities in rural areas can be classified into survival-led or opportunity-led. Survival-led diversification would decrease inequality by increasing the incomes of poorer households and thus reduce poverty.

By an opportunity-led diversification would increase inequality and has a minor effect on poverty, as it tends to be confined to non-poor households. Using data from Western Kenya, we confirm the existence of the differently motivated diversification strategies. Yet, the poverty and inequality implications differ somewhat from the expectations.

Our findings indicate that in addition to asset constraints, rural households also face limited or relatively risky high return opportunities outside the agriculture.

3.4.3 Effects of Modern Agriculture

- Micro nutrient imbalance: Most of the chemical fertilizers used in modern agriculture contain nitrogen phosphorus and potassium (N.R.K) which are macro nutrients when excess of the fertilizers are used in the fields, it causes micro nutrient imbalance.

- Blue Baby syndrome (Nitrate Pollution): When the Nitrogenous fertilizers are applied in the they leach deep into the soil and contaminate the ground water. The nitrate concentration in the water gets increased. When the concentration exceeds 25 mg 1 lit, they cause serious health problem called Blue Baby syndrome. This disease mainly affects infants and can even lead to death.

- Eutrophication: A large proportion of N and P fertilizers used in crop fields is washed off by the runoff water and reaches the water bodies causing over nourishment of the lakes. This process is called as Eutrophication.

  Due to eutrophication, the lakes get attacked by algal blooms. These species use up the nutrients rapidly and grow very fast. Since the life time of the global species are less they die quickly and pollute the water, which in turn affect the aquatic life.

3.4.4 Fertilizer and Pesticide Problems

Problems of Pesticides on Modern Agriculture

In order to improve the crop yield, lots of pesticides are used in the agriculture:

- First generation pesticides: Sulfur, arsenic, lead or mercury are used to kill the pests.
- Second generation pesticides: DDT is used to kill the pests. Although these pesticides protect our crops from huge losses due to pests, they produce number of side effects such as:
    - Death of non-target organisms.
    - Producing new pests.
    - Bio magnification.
    - Risk of cancer.

Effect of Pesticides

- It directly acts as Carcinogens.
- It indirectly suppresses the immune system, the soil and contaminant the ground water. The nitrate concentration in the coater gets increased. When the nitrate concentration exceeds 25 mg/lit., they cause serious health problem called, "Blue Baby Syndrome". This disease affects infants and leads even to death.

Derived qualities of an ideal pesticide

- An ideal pesticide must laid only the target species.
- It must be a biodegradable.
- It should not produce new pests.
- It should not produce any toxic pesticide vapor.
- Excessive synthetic pesticide should not be used.
- Chlorinated pesticides and phosphate pesticides are hazardous, so they should not be used.

3.4.5 Water Logging

Water logging refers to saturation of soil with water.

- Problems in water logging: During water logged conditions, the soil get filled with water and the soil air gets depleted. In such conditions the roots of the plants do not get enough air for respiration. So, mechanical strength of the soil decreases and crop yield fairs.

Causes of Water Logging:

- Heavy rain.
- Excessive water supply to the crop lands.
- Poor drainage.

Remedy:

Preventing excessive irrigation, sub surface drainage technology and bio drainage by trees like eucalyptus tree are some method of preventing water logging.

3.4.6 Salinity

The water, not absorbed by the soil, undergo evaporation's leaving behind a thin layer of dissolved salts in the topsoil. The process of accumulation of salts is called salinity of the soil. The salts are characterized by the accumulation of soluble salts like calcium chloride, sodium chloride, sodium sulfate, magnesium chloride, sodium bicarbonates and sodium carbonates. The pH of the water exceeds 8.0.

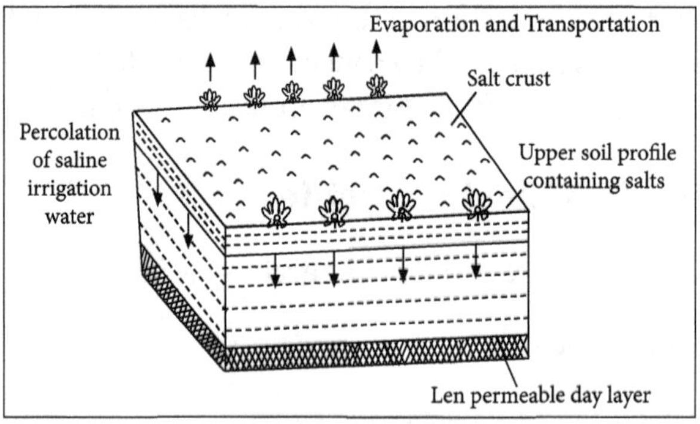

Desalination and Water logging.

Problems in Salinity

Most of the water, used for irrigation comes only from canal or ground, which unlike rainwater contains dissolved salts. Under dry climate or the water gets evaporated leaving behind the salt in the upper portion of the soil. Due to salinity the soil becomes alkaline and crop yield decreases.

Remedy:

- The salt deposit is removed by flushing them out by applying higher grade quality water to such soils.

- Using sub surface drainage system the salt water is flushed out slowly.

3.4.7 Case Studies

Northwest India encompasses two major river basins (the Indus and the Ganges) and lies in the states of Punjab, Haryana and Rajasthan. Although it is a water deficient region, introduction of canal irrigation has reduced the gap between the supply and potential demand to a certain extent. Irrigation is the mainstay of agriculture in this area. As irrigation development took place without parallel development of drainage, water and salt accumulation has occurred in most canal command areas. Salinity has already affected the area of 1 million ha. This area might expand to more than 3 million ha in the next 20 to 30 years unless the remedial measures are taken.

In Punjab and Haryana, surface drains were constructed and groundwater development and flood control were initiated to overcome the waterlogging and salinity problems. In areas where the surface drains do not have a natural drainage outlet, low-head high-discharge pumps dispose the drainage effluent into major canal systems. Horizontal subsurface drainage has been developed on a small scale in the region. Vertical drainage in the form of shallow tube wells is widespread throughout the region. The density of the tube wells varies with the groundwater quality and recharge.

In the areas located in the saline groundwater zones, the annual groundwater recharge from the extensive canal network continues to exceed the groundwater abstraction. As a result, rise in the water table and subsequent salinization has continued in these areas. If national food security is ensured, this problem needs to be addressed. Efforts to date have consisted of lining irrigation channels up to watercourse level and sinking the public tube wells for irrigation. Water testing facilitates have been improved and the technology for the use of tube well water has been disseminated to farmers.

Reuse of saline effluents is an important option for Northwest India as it could supplement the scarce irrigation water supplies and also help to alleviate the disposal problems. Reuse can take place by applying the drainage water directly to the crop, blending it with the canal water and using it intermittently with the canal water. The last form of reuse is most common for private tube well water use as farmers receive canal water only for a few hours per week. The scope for reuse is highest during the winter season when the evaporative demands are low and initial soil salinity is low due to the leaching which occurs during the monsoon rains between the month of June and September. During winter season, soil salinity will increases slowly. When summer starts, the crops are at maturity stage and are able to tolerate the higher levels of salinity. The subsequent monsoon rains will leach the salts that have accumulated during the winter and early summer. If the monsoon rains are not sufficient, heavy pre-irrigation might be given to ensure that the salinity is reduced to acceptable levels for a good germination of subsequent winter crop.

In addition to the drainage water reuse, shallow water table management is an important mechanism for the use of soil water in lands provided with drainage. Experiments showed that on a sandy loam soil with a water table at a depth of 1.7 m and with a salinity of 8.7 dS/m, the water table contributed up to 50% of the water requirement when irrigation was withheld. Similarly, a shallow water table at a depth of about 1.0 m and with a salinity of 3.0-5.5 dS/m facilitated the achievement of potential yields even when the surface water application was reduced to 50%. The salinity buildup was negligible and the small amount of accumulated salts was leached in the subsequent monsoon season.

For the maintenance of a favourable salt balance in the soil and groundwater, the salt outflow from the system should at least equal the salt inflow plus any salt generation in the system itself. For Northwest India, possible methods of disposing of the drainage effluents includes:

- Pumping into the main and branch canals that carry high flow discharges for most of the year.
- Disposal into the regional surface drainage system that links the major rivers flowing through the region.
- Disposal into evaporation ponds.

During the monsoon period, the Yamuna River carries a very high flow discharge. The salinity of the river water during the monsoon is less than 0.2 dS/m. The high flow and low salinity of the water during the monsoon period indicate the potential for the disposal of saline effluents into this river. In Haryana, an area of 1633000 ha drains into the Yamuna River. The extensive surface drainage system to evacuate floodwaters has been constructed. Projections regarding the amount of subsurface drainage water that could be disposed through the surface drains into the Yamuna River were made on the basis of a study covering a period of five years. In the basin, waterlogging and salinity affect 183000 ha. The study shows that the entire amount of subsurface drainage water could be disposed into the Yamuna River during the monsoon period, whereby the river water salinity would remain below 0.75 dS/m. In the winter months, low river flows reduce the disposal capacity considerably. However, in this period, the disposal requirements can be reduced through the implementation of shallow water table management and the reuse of generated drainage water.

As the need for saline water disposal is increasing, evaporation ponds might need to be constructed. The performance of an evaporation pond constructed in a sandy area at Hisar was not encouraging, probably the result of locating the pond in the slightly higher area. It may be necessary to construct a series of ponds in the lowest-lying areas.

It is necessary to maintain a fine balance between reuse and disposal in order to establish a favourable salt regime in the region. The existing experience from small pilot projects is only indicative of the feasibility of reuse, shallow water table management and disposal requirements. The large-scale drainage disposal and reuse programmes planned for Northwest India will need to give due weight to maintaining a favourable salt regime at basin level.

3.5 Energy Resources

Conventional Energy Resources

Coal: Coal is a solid fossil full formed in several stages as buried remains of land plants that lived 300–400 million years ago were subjected to intense heat and pressure over millions of years.

Various Stages of Coal: The various stages of coal during the coalification of wood is:

Wood → Peat → Lignite → Bituminous.

The carbonate content of anthracitic is 90% and its calorific value is 8700 k.cal. The carbon content of bituminous, lignite, pert and wood are 80, 70, 60 and 50% respectively.

Disadvantages:

- When coal is burnt it produces Co2, which causes global warming.
- Since coal contains impurities like S and N, it produces toxic gases during burning:
    - Natural Gas: Natural gas is found above the oil in oil well. It is a mixture of 50-90% methane and a small amount of other hydrocarbons. Its calorific value ranges from 12,000- 14,000 k.cal/m3.
    - Dry Gas: If the natural gas contains lower hydrocarbons like methane and ethane, it is called dry gas.
    - Wet Gas: If the natural gas contains higher hydrocarbons like propane, butane along with methane it is called wet gas.

Like petroleum oil, natural gas can also be formed by the decomposition of dead animals and plant, that were buried under lake and ocean, at high temperature and pressure for million of years.

Tidal Power

Ocean tides, produced by gravitational forces of the sun and moon, contain enormous amount of energy. The 'high tide' and 'low tide' refer to the rise and fall of water in the oceans. The tidal energy can be harnessed by constructing a tidal barrage.

During hide tide, the sea water is allowed to flow into reservoir of the barrage and rotates the turbine, which in turn produces electricity by rotation of the generators.

During the low tide, when the sea level is low, the sea water stored in the barrage reservoir is allowed to flow into the sea and again rotates the turbine.

- Water flows into the reservoir from sea.
- Water flows are from the reservoir to the sea.

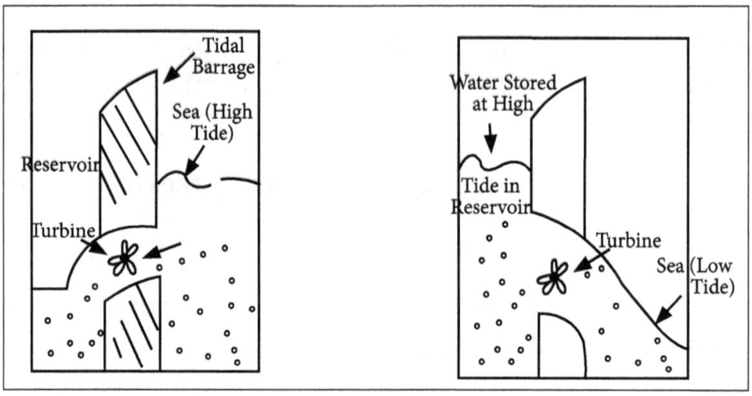

Water flow into and from the reservoir.

Geothermal Energy

The temperature of the earth increases at a rate of 20 - 75°C per km, when we move down the earth surface. High temperature and high pressure steam fields exists below the earth's surface in many places. The energy harnessed from the high temperature present inside the earth is called geothermal energy.

- Natural geysers: In some of the places, hot water or steam comes out of the ground through cracks naturally in the form of natural geysers.
- Artificial geysers: In some of the places, we can artificially dried a hole up to the hot region and by sending a pipe in it, we can make hot water or steam to rush out through the pipe with very high pressure.

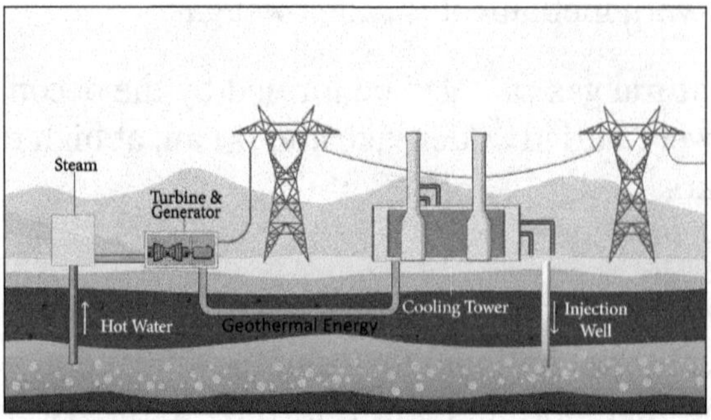

Artificial geysers.

Thus, the hot water or steam coming out from the natural or artificial geysers is allowed to rotate the turbine of a generator to produce electricity.

Biomass Energy

Biomass is the organic matter, produced by plants or animals, used as source of energy; most of the biomass is burned directly for cooling, heating and industrial purposes.

Examples: Wood, crop residues, seeds, cattle dung, sewage, agricultural wastes, etc. Biomass energies are of any one of the following types.

Bio gas: Bio gas is a mixture of gases such as methane, carbon dioxide, hydrogen sulfide, etc. It contains about 65% of methane gas as a major constituent. Bio gas is obtained by the anaerobic fermentation of animal dung or plant wastes in the presence of water.

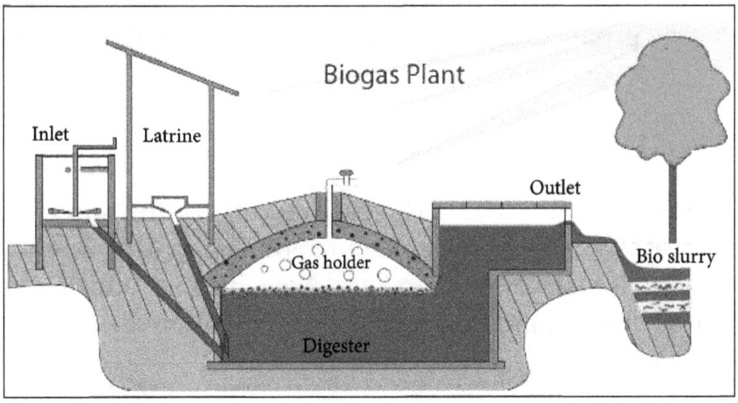

Biogas plant.

Bio fuels: Bio fuels are the fuels, obtained by the fermentation of bio mass. Examples: Ethanol, Methanol.

- Ethanol: Ethanol can be produced from sugarcane. Its calorific value is less when compared to petrol and produces much less heat than petrol.
- Methanol: Methanol can be obtained from ethanol or sugar containing plants. Its calorific value is also too low when compared to diesel and gasoline.
- Gasohol: Gasohol is a mixture of ethanol and gasoline. In India, a trial is being carried out to use Gasohol in cars and buses.

Hydrogen fuel: Hydrogen can be produced by thermal dissociation or photolysis or electrolysis of water. It possess high calorific value. It is nonpolluting, because the combustion predict is water.

$$2H_2 + O_2 \rightarrow 2H_2O + 150$$

Disadvantages of hydrogen fuel:

- Safe handling is required.
- Hydrogen is difficult to store and transport.
- Hydrogen is highly inflammable and explosive in nature.

3.5.1 Growing Energy Needs

Sources of Energy

Energy may be defined as, "any property, which can be converted into work".

Energy is available on the earth in a number of forms, some of which may be immediately used to do work, while the others require some process of transformation. All the development activities in world are directly or indirectly dependent upon energy. Both energy production and energy utilization are the indicators of a country's progress.

Development of Energy

First form of energy is the fire. The early mass discovered fire and used it for cooling and heating purposes. Wood is the main source of energy, which is later replaced by coal. Coal is new being replaced by oil and gas. Now due to insufficient availability and price hike, people started of thinking and using several alternate sources of the energy.

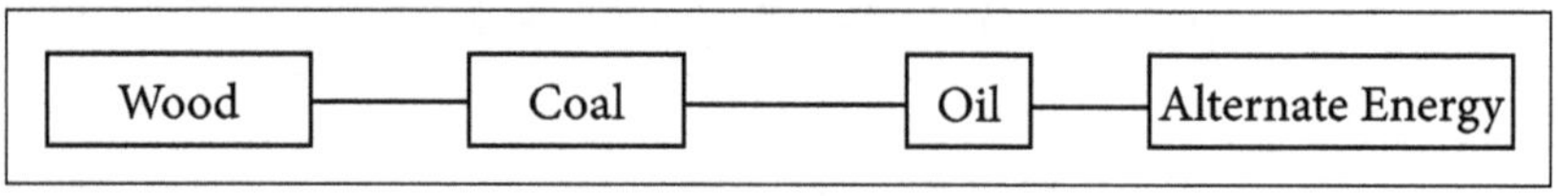

Sourcing Energy Needs

Energy is essential to all human societies. All industrial process like lighting, mining, transport, heating and cooling in buildings, all require energy. With the demands of the growing population, the world is facing further energy deficit.

Our life style is also changing from sample way of life to luxurious life style. At present 95% of the commercial energy is available only from the fossils fuels like oil, coal and natural gas are not going to last for many more years. It would be really ironic if fuel become more expensive than the food.

Energy Distribution World Scenario

Developed countries like U.S.A and Canada constitute only 5% of the world's population, but consume 25% of the available world's energy resources. It has been observed, that in USA and Canada an average person consumer 300GJ per year.

So a person in a developed country consumes almost as much energy in a single day as one person consumes in a whole year in a poor country. From the scenario above, it is clear that own life style and standard of living are closely related to energy needs.

Per capita Energy Use

The correlation is obtained between per capita energy use and GNP. The developed countries like U.S.A., Japan, Switzerland, etc., with help GNP show high energy use while poor countries like India, China, Ethiopia with GNP show low energy use.

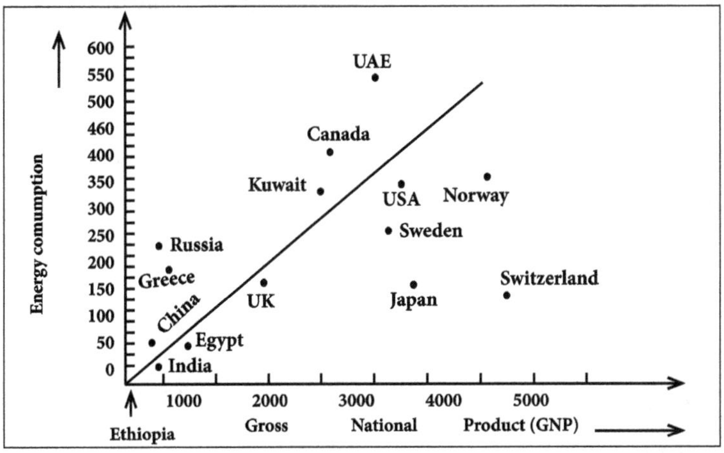

Per capita Energy use.

3.5.2 Renewable and Non-Renewable Energy Sources

Renewable Energy Resources

Solar Energy: The energy that we get directly from the sun is called solar energy. The nuclear fusion reactions occurring inside the sun release enormous amount of energy in the form of heat and light. Various techniques are available for collecting, converting and using the solar energy.

Methods of Harvesting Solar Energy

Some important solar energy, harvesting devices are given below:

Solar Cells

Solar cells consists of a p-type semiconductor and n-type semiconductor. The are in close contact with each other. When solar rays fall on the top layer of p-type semiconductor. The electrons from the valence band get promoted to the conduction band and cross the p-n junctions into n-type semiconductor, thereby potential difference between two layers is created which causes flow of electrons.

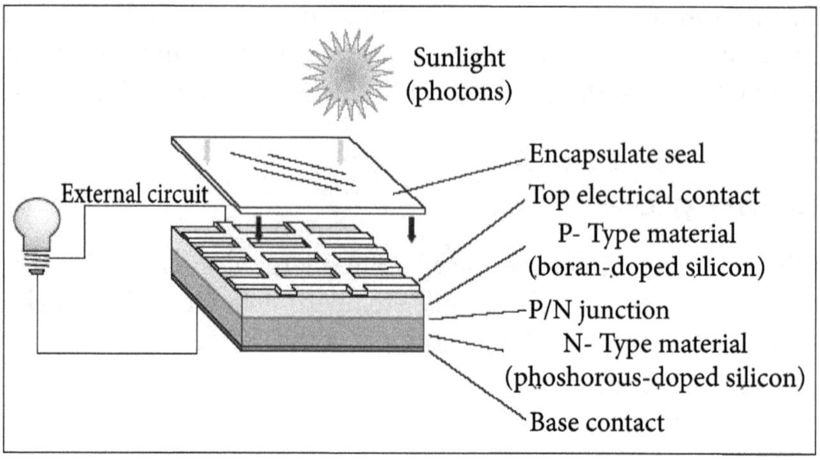

Solar cell.

Uses: Solar cells are used in electronic watches, calculators, water pumps, street lights and to run radio and TVs.

Solar Battery

When a large number of solar cells are connected in series it forms a solar battery. Solar battery produces electricity capable of running water pump to run street-light etc. They are used in remote areas where conventional electricity supply is a problem.

Solar Water Heater

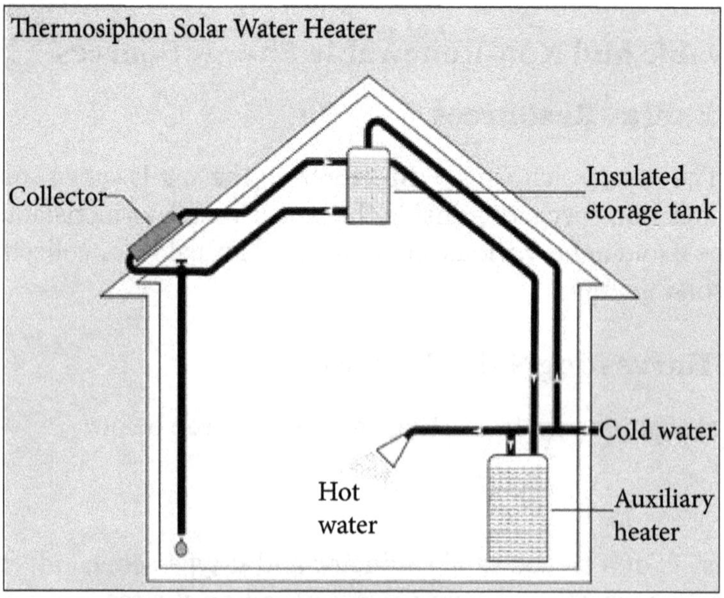

Solar water heater.

It consists of an insulated box inside of which is painted with black paint. It is also provided with a glass lid to receive and store the solar heat. Inside the box, it has a black painted copper coil, through which cold water is allowed to flow in, which gets heated up and flows out into a storage tank. From the storage tank water is then supplied through pipes.

Significance of solar energy:

- Solar water heaters, cookers, require neither fuel not attention while cooking the food.
- Solar cells are free from air and sound pollution.
- Solar cells can be used in remote and isolated areas, forests, hilly regions.

Solar Heat Collectors

Solar heat collectors consists of the natural materials like stones, bricks or materials like glass, which can absorb heat during the day time and release it slowly at night.

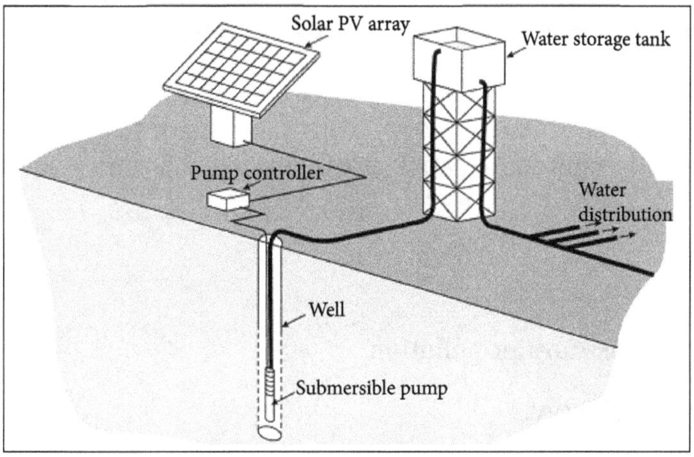

Solar pump run by solar cells

Uses: It is generally used in cold places, where the houses are kept in hot condition using solar heat collectors.

Wind Energy

Moving air is called wind. Energy covered from the force of the wind is called wind energy. The energy possessed by wind is because of its high speed. The wind energy is harnessed by making use of wind mills.

Wind Mills

The strike on the blades of the wind will make it rotating continuously. The rotational motion of blades drives a number of machines like water pump, flour mills and electric generators.

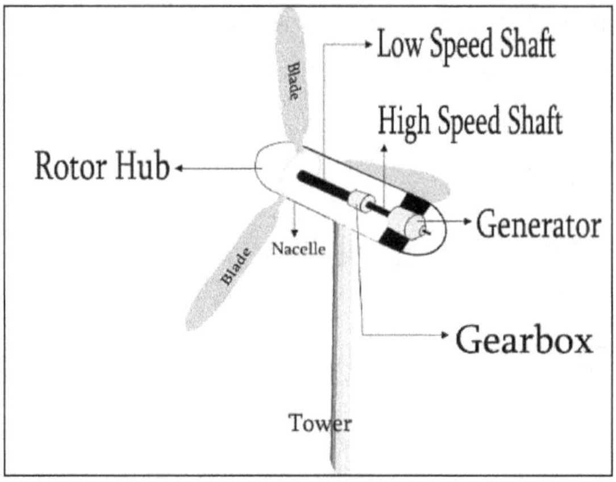

Wind mill.

Wind Forms

When a large number of wind mills are installed and joined together in definite pattern it forms a wind farm. The wind farms produce large amount of electricity.

Condition

The minimum speed required for satisfactory working of a wind generator is 15 km/hr.

Advantages:

- It is very cheap.
- It does not cause any air pollution.

Significance of Wind Energy:

- It is made available easily in many off-shore, on-shore and remote areas.
- The generation period of wind energy is low and power generation starts from commissioning.
- It is recommended to broaden the nation's energy options for new energy sources.

Ocean Energy

Ocean can also be used for generating energy in the following ways.

Ocean Thermal Energy

There is often a large temperature difference between the surface level and deeper level of the tropical oceans. This temperature difference can be utilized to generate electricity. The energy available due to the difference in temperature of water is called thermal energy.

Condition

The temperature difference should be 20°C or more is required between surface water and ground water.

Process

The warm surface water of ocean is used to boil a low boiling liquid like ammonia. The high vapor pressure of the generator and generates electricity. The cold water from the deeper ocean is pumped to cool and condense the vapor into liquid.

Significance of OTE

- OTE is continuous, renewable and pollution free.
- The use of cold deep water as the chiller fluid in air-conditioning has also been proposed.
- Electric power generated by OTE can be used to produce hydrogen.

Geo-Thermal Energy

- The power generation level is higher for geothermal than for solar and wind energies.
- Geothermal power plants can be brought on line more quickly than most other energy sources.
- GTE is effectively and efficiently used for direct uses such as hot water bath, resorts and aquaculture green houses.

Tidal Energy or Tidal Power

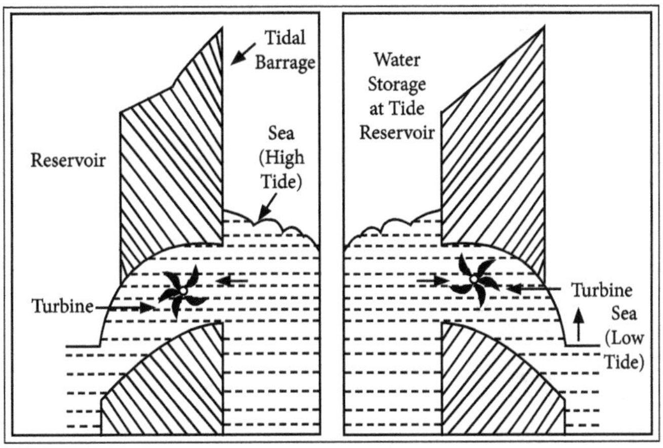

(a) Water flows into the reservoir from sea (b) Water flows out from the reservoir to the sea.

Ocean tides, produced by gravitational forces of sun and moon, contain enormous amount of energy. The "high tide" and "low tide" refer can be to the rise and fall of water in the oceans. The tidal energy shall be harnessed by constructing a tidal barrage.

During high tide, the seawater is allowed to flow into the reservoir of barrage and rotates the turbine which in turn produces electricity by rotating the generators.

During low tide, when the sea level is low, the sea water stored in the barrage reservoir is allowed to flow in and in turn rotating the turbine again.

Significance of Tidal Energy

- Tidal power plants do not require large areas of valuable lands as they are on the bays or estuaries.
- As the sea water is inexhaustible, it is completely independent of the uncertainty of precipitations.
- It is pollution free energy source as it does not use any fuel and also does not produce any wastes.

Tidal Energy

1. Natural Geysers: In some places, hot water or steam comes out of the ground through cracks naturally in the form of natural geysers.

2. Artificial Geysers: In some places, we can artificially drill a hole up to the hot region and by sending a pipe in it; we can make the hot water or steam to rush out through the pipe with very high pressure. Thus, hot water coming out from the natural or artificial geysers is allowed to rotate the turbine of a generator to produce electricity.

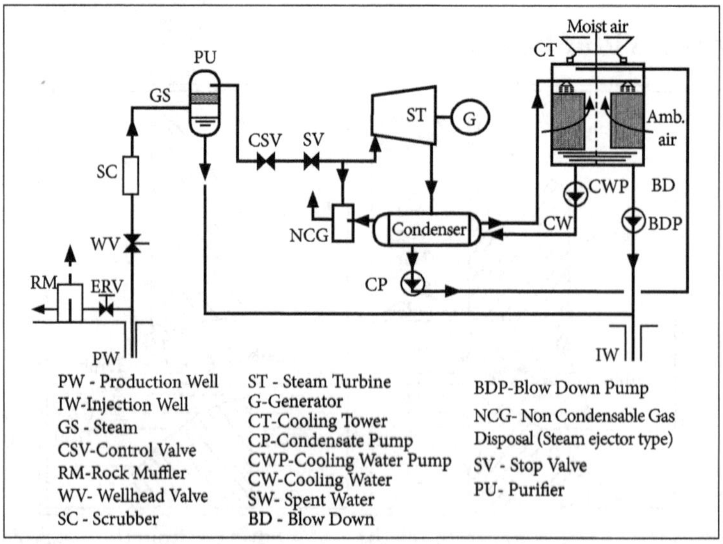

Geothermal power plants.

Significance of Geo-Thermal Energy

- The power generation level is higher for geothermal than for solar and wind energies.
- Geothermal power plants can be brought on line more quickly than most other energy source.
- GTE is effectively and efficiently used for direct uses such as hot water bath, resorts, aquaculture, green houses.

Biomass Energy

Biomass is the organic matter, produced by plants or animals used as sources of energy. Most of the biomass is burned directly for heating and cooling and industrial purpose.

Examples: Wood, crop residues, seeds cattle dung, etc.

3.5.3 Use of Alternate Energy Sources

While energy surrounds us in all aspects of life, the ability to harness it and use it for constructive ends as economically as possible is the challenge before mankind.

Fossil fuels have become essential for human society. Their consumption in terms of coal equivalent was about 1 billion tons in 1900 A.D and was about 3 billion metric tons in 1950 A.D. It is now l5 billion metric tons in the year 2000. The convenience of oil and its high energy content makes it the most important fuel today.

These fossil fuels are 'non-renewable' resources. Over consumption of these resources shall naturally cause their depletion which result in a chain of reactions culminating into drastic changes in economic, pattern of agriculture and even our daily life. Combustion of fossil fuels which yield energy also produces numerous of pollutants. Major pollutants produced are: Carbon dioxide, Carbon monoxide, Sulfur dioxide, Oxides of nitrogen, hydrocarbons and particulates.

Alternative energy refers to energy sources which are not based on the burning of fossil fuels or the splitting of atoms. The renewed interest in this field of study comes from undesirable effects of pollution both from burning fossil fuels and from nuclear waste byproducts.

Fortunately there are several means of harnessing energy which have less damaging impacts on our environment. The alternate energy should be convenient, clean and less wasteful of energy and should be renewable so that there is no threat of its depletion.

Solar energy is one the most resourceful sources of energy for the future. One of the reasons for this is that the total energy we receive each year from the sun is around 35,000 times the total energy used by man. However, about 1/3rd of this energy is either absorbed by the outer atmosphere or reflected back into space. Solar energy is presently being used on smaller scale in furnaces for homes and to heat up swimming pools. On a larger scale use, solar energy can be used to run cars, power plants and space ships.

Wind power is another alternative energy source that can be used without producing by products that are harmful to nature. Like solar power, harnessing the wind is highly dependent upon weather and location. The average wind velocity of Earth is around 9 m/sec. And the power that can be produced when a wind mill is facing the wind of 10 mi/hr is around 50 watts.

Geothermal energy is another energy source, although it is not resourceful enough to replace more than a minor amount of future's energy needs. Geothermal energy is obtained from the internal heat of the planet and can be used to generate steam to run a steam turbine. This in turn generates electricity, which is a very useful form of energy.

Some systems pump hot water into permeable sedimentary hot pots found underground and then use the steam to generate electricity. Then the used steam is condensed and sent back down to the permeable sedimentary stream. Another system utilizes volcanic magma which is still partly molten at around 650 degrees, Celsius, to boil water which would generate electricity. There is also a system which uses hot dry rock, but still it is extremely hot.

To recover this heat from the rocks, a system is used which circulates water through the rock and transfers the heat up to a steam generator. The first system listed here is not as useful as other methods because of the acidic nature of fluids that are found under the ground. These acidities require a lot of maintenance and upkeep on the equipment and its cost reduces the economic effectiveness of the system. Hence, geothermal energy systems are more inefficient than the other alternative energy sources because of the costs required in upkeep and the shortage of potential sites.

Tidal power, sometimes known as the tidal energy, is a form of hydro power that exploits the rise and fall in sea levels due to the tides or the movement of water caused by the tidal flow. Because the tidal forces are caused by interaction between gravity of the Earth, Moon and Sun, tidal power is essentially inexhaustible and classified as a renewable energy source.

In fact though, the ultimate energy source is the rotational energy of the Barth, which will not run out in next 4 billion years, although the Earth's oceans may boil away in 2 billion years. Although not yet widely used, tidal power has great potential for future electricity generation and is more predictable than wind energy and solar power. In Europe, the tide mills have been used for nearly a thousand years, mainly for grinding grains.

Hydroelectricity comes from the damming of the rivers and by utilizing the potential energy stored in the water. As the water stored behind a dam is released at high pressure, its kinetic energy is transferred onto turbine blades and used to generate electricity. The system has enormous costs up front, but has relatively low maintenance costs and provides power quite cheaply. In the United States approximately 180,000 MW of hydroelectric power potential is available and about a third of that is currently being harnessed.

3.5.4 Energy Conversion Process

Energy transformation or energy conversion is the process of changing one form of energy into another form of energy. In physics, the term energy describes the capacity to produce certain changes within any system, without regard to limitations in transformation imposed. Changes in total energy of systems can only be accomplished by adding or removing energy from them, as energy is a quantity which is conserved, as stated by the first law of thermodynamics. Mass-energy equivalence, which rose up from special relativity, states that changes in the energy of systems will also coincide with changes in the system's mass and the mass of a system is a measure of its energy content.

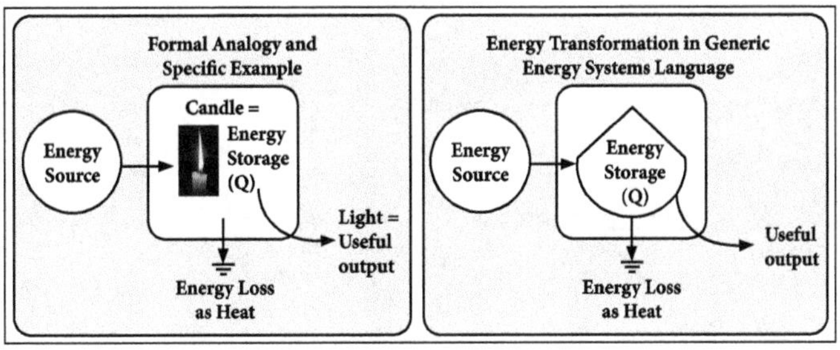

$$\text{Energy\%} = \frac{\text{Useful Output}}{\text{Total Input from Source}} \times \frac{100}{1}$$

Energy in many of its forms may be used in natural processes or to provide some service to society such as heating, refrigeration, light or performing mechanical energy to operate machines. For example, an internal combustion engine converts the potential chemical energy in gasoline and oxygen into thermal energy which is transformed into the mechanical energy that accelerates the vehicle. A thermal or heat energy converts the chemical energy into mechanical energy that can be used.

3.5.5 Biogas: Production and Uses

Bio gas is produced from plant material and animal waste, garbage, waste from the households and some types of industrial wastes such as dairies, fish processing and sewage treatment plants. It is a mixture of gases that includes methane, carbon dioxide, hydrogen sulfide and water vapor. In this mixture, methane burns easily.

Generally, it is believed that with one ton food waste, one can produce 85 Cu. M of bio gas. Once used, the residue can be used as an agricultural fertilizer. Denmark produces a large quantity of the biogas from the waste and produces around 15,000 megawatts of electricity from 15 farmers cooperatives. London has a plant which can make 30 megawatts of electricity a year by using 420,000 tons of municipal waste which gives power to 50,000 families. In Germany, 25% of the landfills for garbage produce power from bio gas. Japan uses 85% of its waste and France about 50%. Bio gas plants have become increasingly popular in India in the rural sector.

The bio gas plants use cow dung, which is converted into a gas which is used as a fuel. It is also used for running dual fuel engines. The reduction in kitchen smoke by using bio gas has reduced lung conditions in thousands of homes. The fibrous waste of the sugar industry is the world's largest potential source of the biomass energy.

Ethanol produced from sugarcane molasses is a good automobile fuel and is now used in a third of the vehicles in Brazil. The National Project on Bio gas Development (NPBD) and Community. Institutional Bio gas Plant Program promotes various bio gas projects. By 1996 there were already 2.18 million families in India that used bio gas. However, China has 20 million households using bio gas.

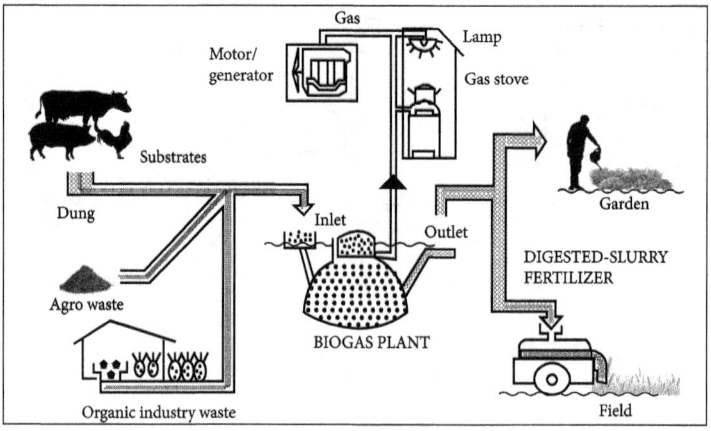

Biogas production.

3.5.6 Anaerobic Digestion

Anaerobic digestion is the natural breakdown of organic materials into methane and carbon di oxide gas and fertilizer. This takes place naturally or in the anaerobic digester.

A typical anaerobic digester is a sealed vessel in which bacteria act without oxygen. The organic material contents need to be fully mixed and warmed, usually to the blood temperature.

Biogas is the mixture of gases formed during the anaerobic digestion of organic wastes. Biogas consists of methane (c70%) and carbon dioxide (c30%). It is used in the stationary engines to generate electricity, but it is not suitable as a vehicle fuel after removing the carbon dioxide, the remaining methane is known as Renewable Natural Gas or Biomethane.

Biomethane is virtually identical to natural gas, the main difference being that it is produced in days, rather than taking millions of years, billions of years ago. The uses for biomethane are varied as are those for natural gas, for heating, cooling, as a source of chemicals, fertilizer or hydrogen.

When used as a vehicle fuel, biomethane is the worlds cleanest and the most environmentally friendly fuel.

Carbon dioxide is valued for its properties as an inert gas, for heat transfer and as a solvent. Common uses are as follows:

- Carbonation of soft drinks.
- Fire extinguisher systems.
- Decaffeination of coffee.
- Freezing or chilling of food products.
- Oil extraction; pumped into an oil well it dissolves the oil.
- Shield gas in welding.
- Propellant in aerosol cans.
- Respiration stimulant for medical use.
- In its supercritical form, as the most environmentally friendly solvent.
- Food production in greenhouses.

Compost produced from a thorough and complete treatment system such as the Maltin System ensures that all the material is broken down into its most basic chemical components with:

- No smell when dry or when wetted.
- No contaminants.
- No weed seeds.
- No pathogens.

At organic power, we maintain that WASTE is merely the result of the poor or incomplete processing. A properly designed process will not produce any waste.

3.5.7 Case Studies

Lepelsfontein is a small village in the Northern Cape, some 500 km north of Cape Town in South Africa. It has a population of roughly 450 people. The physical infrastructure comprises gravel roads, manual telephones and a primary school. The water supply system in the Lepelsfontein area comprises a number of sources. The main source is groundwater from a borehole. A three-phase PV submersible centrifugal pump is used to pump water. There is also a diesel pump which is used as a back up to the PV pump in the event of PV pump failure or in times of extended cloud cover. The PV system is situated about 2 km to the south west of Lepelsfontein village. The PV pump operates seven days a week.

A diesel pump becomes operational only when PV pump is not working. Water is pumped to 2 storage tanks located in the village. Each storage tank has a capacity of 30 kl. Water from storage tanks is gravitated to a diesel pump located at about 300 m down the hill. Purifying chemicals are added before the water is pumped to the supply tanks. Water has to be desalinated as it is being claimed that the sea, which is about 10 km to the south east of the settlement, affects the quality of underground water and therefore, it is not readily available for consumption. There are four supply tanks of 10 kl each located at about 400 m from the diesel desalinating pump house uphill. Water from the supply tank is gravitated to the village and is accessed through standpipes. The PV water pumping facility at Lepelsfontein comprises three subsystems namely, a set of PV panels, an inverter and a three-phase submersible centrifugal pump.

Field Results

The results presented in this section are obtained from the Lepelsfontein PV water pumping system. The data logger captured the information from the 03/12/04 to 02/02/05. The results shown below are the results captured only when the pump was in operation.

Array Efficiency vs. Irradiance

Efficiency of an array is the amount of solar energy the module can convert into electrical energy. Figure shows the efficiency of the array at different times of the trial period. The array efficiency is dependent on several factors such as; time of day, weather conditions, wind speed and irradiance as well as the temperature of the array. The maximum array efficiency obtained was 3.9%. The obtained efficiency of 3.0% is quite low and the discrepancy of the results at high solar intensities is attributed to the temperature effects of solar cells at the elevated cell temperatures. These high temperatures increase the irradiance power thus having a negative impact on array power generation. If the assumed optimum efficiency point of the pump is considered, the expected array efficiency of 22% can be taken, which will affect the overall system efficiency proportionately. At the operating condition of 2.6 m^3/day, the efficiency is as low as 15%.

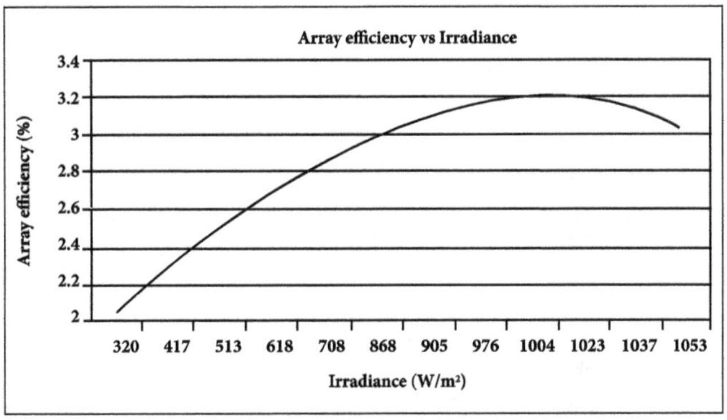

Array efficiency versus irradiance.

Overall Efficiency Vs. Time

The overall efficiency of the system is determined by the power delivered by the pump and power from the sun. Figure shows the efficiencies ranging to a maximum of 3.2%, which is acceptable when one considers an overall system efficiency of 3% and 4% from the Intermediate Technology Development Group Manual and other literature.

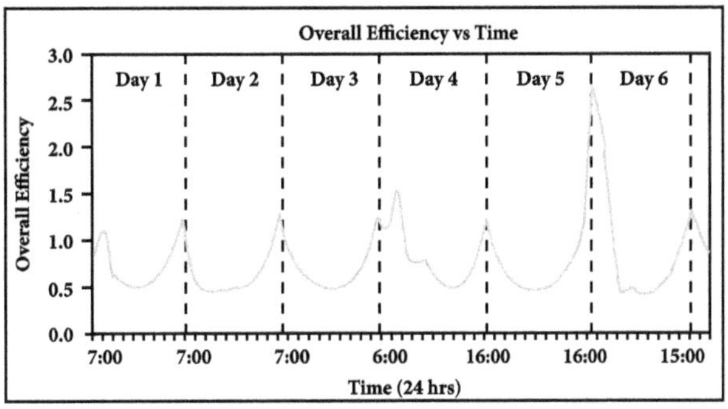

Overall efficiency versus time.

Table: Summary of Field Results.

Component	Array	System	Overall
Actual efficiency	4.0 %	32.0%	1.3%
Design efficiency	20.0%	30-40&	6-8%

Economic Results and Analysis

To establish whether this PV pumping system is economically viable when it is compared with other pumping technologies, the results from this system were used to perform a life cycle costing and a comparison was made with a comparably-sized diesel system. A 15, 20, 25 year life-cycle costing of the field PV and diesel systems was performed and the actual results are presented in the table is also a prediction of unit (R/

m^3) and pumping costs (R/m^4) if design conditions were achieved, i.e., flow of 20m³ / day. Figure below shows the comparison of PV actual and diesel systems cumulative and energy costs.

Pv Actual and Diesel Water Pumping Costs

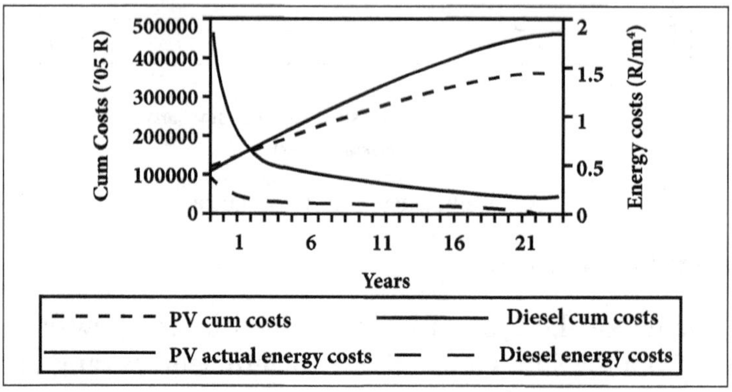

PV actual & diesel systems cumulative & energy costs.

The life cycle cost analysis of the PV pumping system indicates that the average pumping costs over 25 years are likely to be below 22 cents/m^4, while pumping costs of diesel are likely to be below 7 cents/m4 when the pump operates at 3.6m3 /day.

Pv Predicted and Diesel Water Pumping Costs

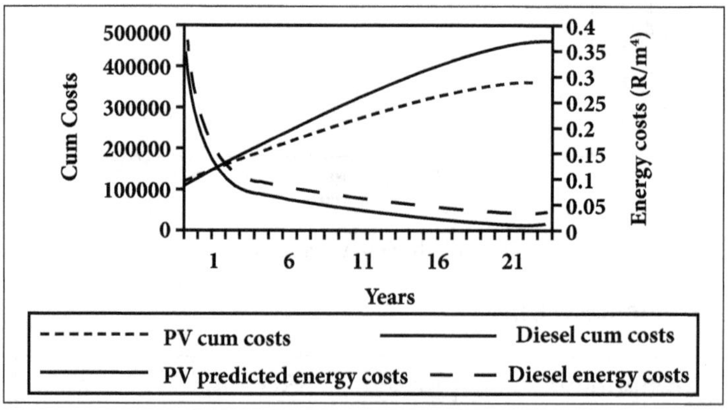

However, if the pump operated at its optimum design conditions (20m³ /day), then the pumping costs of PV would reduce to 4 cents/m^4, 43% less than that of diesel. From the two water-pumping technologies, the unit water cost for Lepelsfontein is about five times the unit cost of Rooifontein. These could be caused by a number of factors, including a high capital cost for Lepelsfontein, very low pump output as well as high operating costs. The pumping costs for Lepelsfontein are about 3 times higher than those of Rooifontein. In contrast, the unit water cost is 90% that of diesel and the pumping cost is about 60% that of diesel when the design conditions for PV are assumed.

Summary of Pv, Diesel Life Cycle (Lcc), Unit (Uwc) and Pumping Costs (Ec)

Pv						Diesel		
yrs	Lcc(R)	UWC 3.6m³/day	UWC 20m³/day	EC 3.6m³/day	EC 3.6m³/day	LCC (R)	UWC 21m³/day	EC 21m³/day
15	293207	R15.08/m³	R2.72/m³	R0.302/m⁴	R0.054/m⁴	358310	R3.12/m³	R0.087/m⁴
20	331390	R12.79/m³	R2.30/m³	R0.256/m⁴	R0.046/m⁴	413656	R2.70/m³	R0.075/m⁴
25	361024	R1114/m³	R2.01/m³	R0.223/m⁴	R0.040/m⁴	456611	R2.38/m³	R0.066/m⁴

The findings of this study do support the existing body of evidence, which indicates that PV pumping can be competitive with diesel water pumping under specific head and flow conditions. However, the results obtained in this study are short-term results, making it difficult to make valid judgments with regard to how this system behaves in the long run. Yet the socio-institutional implementation strategies are crucial to the techno-economic success of actual pumping schemes. Even if the diesel generators may appear to be cheaper on a life cycle cost comparison, it might be preferable to opt for a PV system because of the operational advantages. Future work will evaluate a larger number of systems and eventually record long-term results which will be used to assess the reliability and functionality of the systems.

3.6 Land Resources

The occupation is nearly 20 % of the earth surface, measuring almost up to about 13000 million hectares of area. The houses, roads and factories occupy nearly one third of the land. The forests occupy another one third of the land.

The rest of land is used for and for meadows and pastures. The soil forms the surface layer of the land which covers more than the 80 % of land. The soil is defined as a natural body which keeps on changing and even allows the plants to grow.

It consists of organic and inorganic materials. This definition is given by the Buck man and Brady. The branch of science which can deal with the formation and distribution of soil in the different parts of the world is referred as a pedology. The professional which deals with the soil is known as the pedologist.

The inorganic component in the soil is 45 % and the organic component in the soil is 5 %. The water component in the soil is 25 % and air component in the soil is 25 %. The soil particles have fine spaces which are known as the pore spaces. These are also known as the interstices. They contain air and water along with the dissolved substances. The water and air content in the soil is inversely related to each other. The more is the water content lesser is the space for air to exist. The soil has both animals and plants. The micro flora consists of the heterotrophic and autotrophic bacteria. It also contains the fungi and algae. The heterotrophic bacterium consists of nitrogen and non-nitrogen fixing bacteria. The nitrogen fixing bacteria can be symbiotic, non-symbiotic, aerobic and anaerobic. The non-nitrogen fixing bacteria can be either aerobic or anaerobic. The fungus includes the yeast and mushrooms.

The algae can be red or brown or green. The fauna can be micro or macro. The micro fauna includes the protozoa and nematodes. The macro fauna includes the mites, termites, earthworm, snails and mice. The soil has different types of soil particles. The mineral composition of the rock determines them along with the size of particles. It includes gravel particles, sand, silt and clay particles. The gravel particles consist of mainly small stones and have a few sand particles and are used to make roads. The sand particles consist of pores and are aerated. They can hold a little bit of water and are made up of large quartz. The silt particles are moved by the help of water. They are left at the bank of river. They are inert and are made up of large quartz.

The clay particles contains nutritive salts and have ability to retain the water. They are not inert and react chemically. Some of their pure forms are not suitable for the growth of plants as they form a non-penetrable mass. The other components of soil mix with the clay particle and form a granular soil. This type of soil is ideal for the cultivation. It contains pores as well as has the ability to hold water. It also contains nutritive salts.

The loamy soil is made up of clay, silt and sand. The proportion of the clay is least and is half as compared to the silt and sand. The silt and the sand are twice and equal in the proportion. It is also a good soil for the growth of plants as it has pores as well as has the ability to hold water. It also contains some nutritive salts.

There are many factors which control the nature of soil. They are porosity, water holding capacity and texture. They come under the physical nature of soil. The chemical nature of the soil is governed by the salt content, inorganic and organic content includes certain metals. The climate, topography and the organisms also play a vital role in deciding the nature of soil. The half decayed and half synthesized part of organic material in the soil forms the humus. It contains nutrients and help in growth. It has the ability to absorb the heat and warm the soil. It makes the soil granular by its porosity and water holding capacity.

3.6.1 Land Degradation

Land degradation is the process of deterioration of soil or loss of fertility of the soil. It is the reduction in capacity of the land to provide ecosystem goods and services and assure its functions over a period of time. Land degradation affects large areas and many people in the dry land regions. Increased population pressures and excessive human expansion into dry lands during long wet periods leave an increasing number of people stranded there during dry periods. The transfer of critical production elements to other uses through the introduction of irrigated and non-irrigated cash crops and the use of water for the industrial and urban purposes at the expense of rural agricultural producer break links in the traditional production chains in dry lands. The removal of protective cover to reduce the competition for ploughing, water and nutrients, heavy grazing and deforestation leave the soil highly vulnerable to wind erosion particularly during severe droughts. Heavy grazing around water points or during long droughts prevents the regrowth of vegetation or favors only unpalatable shrubs.

Harmful effects of land (soil) degradation:

- The soil texture and soil structure are deteriorated.
- Increase in water logging, salinity, alkalinity and acidity problems.
- Loss of soil fertility, due to loss of invaluable nutrients.
- Loss of economic social and biodiversity.

Causes of Land Degradation

- Population: As population increases, more land is needed, for producing food, fiber and fuel wood. Hence there is more and more pressure for the limited land resources, which are getting degraded due to over exploitation.
- Urbanization: The increased urbanization due to population growth reduces extent of agricultural land. To compensate the loss of agricultural land, new lands comprising natural ecosystems such as forests are cleared. Thus the urbanization leads to deforestation, which in turn affects millions of plant and animal species.
- Fertilizers and Pesticides: Increased applications of fertilizers and pesticides are needed to increase farm output in the new lands, which again leads to pollution of land and water and soil degradation.
- Damage of Top Soil: Increase in soil food production generally leads to damage of top soil through nutrient depletion.
- Water Logging, soil erosion, desalination and contamination of the soil with Industrial wastes all cause land degradation.

3.6.2 Man Induced Landslides

The causes of man induced landslides are as follows:

- Excavation,

- Loading,
- Draw-down,
- Land use (e.g. construction of roads, houses etc.),
- Water management,
- Mining,
- Quarrying,
- Vibration,
- Water leakage,
- Deforestation,
- Land use pattern,
- Pollution.

3.6.3 Soil Erosion and Desertification

Soil is naturally created when the small pieces of weathered rocks and minerals mix with the organic materials from decaying plants and animals. Soil creation is a slow process, taking many years. However, the soil that is created is constantly subjected to natural and manmade forces that disrupt it.

Soil erosion is a naturally occurring process that affects all the landforms. In agriculture, soil erosion refers to the wearing a way of field's topsoil by the natural physical forces of water and wind or through the forces associated with farming activities such as tillage.

Causes of Soil Erosion

Erosion occurs when farming practices are not compatible with the fact that soil can be washed away or blown away. These practices are as follows:

- Inappropriate farming techniques such as deep ploughing of land 2 or 3 times per year to produce the annual crops.
- Overstocking and overgrazing.
- Planting the crops down the contour instead of along it.
- Lack of crop rotation.

Erosion, whether it is by water, wind or tillage, involves three distinct actions such as soil detachment, movement and deposition. Topsoil, which is high in organic matter, fertility and soil life is relocated elsewhere "on-site" where it builds up over time or is carried "off-site" where it fills in the drainage channels. Soil erosion reduces the cropland productivity and contributes to the pollution of adjacent wetlands, watercourses and lakes.

Soil erosion can be a slow process that continues relatively unnoticed or can occur at an alarming rate, causing serious loss of topsoil. Soil compaction, loss of soil structure, low organic matter, salinisation, poor internal drainage and soil acidity problems are other serious soil degradation conditions that can accelerate the soil erosion process.

Soil erosion is defined as the wearing away of the topsoil. Topsoil is the top layer of soil and is the most fertile because it contains the most organic, nutrient-rich materials. Hence, this is the layer that farmers wants to protect for growing their crops and ranchers want to protect for growing grasses for their cattle to graze on.

One of the main causes of soil erosion is water erosion, which is the loss of topsoil due to water. Raindrops fall directly on the topsoil. The impact of the raindrops loosens the material bonding it together, allowing small fragments to detach. If the rainfall continues, water gathers on the ground, causing water flow on the land surface which is termed as surface water runoff. This runoff carries the detached soil materials away and deposits them elsewhere.

There are some conditions that can accentuate the surface water runoff and therefore, soil erosion. For example, if the land is sloped, there is a greater potential for soil erosion due to gravity which pulls the water and soil materials down the slope.

Effects of Soil Erosion

The loss of natural nutrients and possible fertilizers directly affect the crop growth, emergence and yield. Seeds can be disturbed or removed and pesticides can be carried off. The soil structure, quality, texture and stability are also affected, which in turn affects the holding capacity of the soil.

Eroded soil can inhibit the growth of seeds, bury seedlings, contribute to road damage and even contaminate the water sources and recreational areas.

Desertification

Desertification is a progressive destruction or degradation of arid or semiarid lands to desert. It is also a form of land degradation. Desertification loads to the conversion of range lands or irrigated croplands to desert like conditions in which agricultural productivity falls. Desertification is characterized by denegation, depletion of ground water, desalination and soil erosion.

Causes of Desertification

- Deforestation: The process of deserting and degrading a forest land initiates a desert. If there is no vegetation to hold back the rain water, soil cannot soak and ground water levels do not increases. This also increases, soil erosion, loss of fertility.
- Overgrazing: The increase in cattle population heavily grazes the grass land or forests and as a result denudes the land area. The denuded land becomes dry, loose and more prone to soil erosion and leads to desert.
- Pollution: Excessive use of pesticides and fertilizers and disposal of toxic water into the land also leads to desertification.

Harmful Effects of Desertification

- Around 80% of the productive land in the arid and semiarid regions are converted into desert.
- Around 600 million people are threatened by desertification.

3.6.4 Role of an Individual in Conservation of Natural Resources

Since resources are being exhaustible, it is the duty of every individual on this earth has to conserve the natural resources in such a way that they must be available for future generation also. Individual must understand the essential if natural resources. Due to advancement in technology and population growth, the present world is feeling lot of problems on degradation of natural resources.

Measures Recommended for Conservation of Natural Resources

1. Conservation of Energy

- Switch off lights, fans and other appliances when not in use.
- Use solar heater for cooking our food on sunny days, which will cut down our LPG expenses.
- Dry the clothes in sunlight instead of driers.
- Grow trees near the houses and get a cool breeze and shade. This will cutoff our electricity charges on A/C and waters.
- Use always pressure cooker.
- Ride bi-cycle or just walk instead of using car and scooter.

2. Conservation of Water

- Check for water leaks in pipes and toilet and repair them promptly.

- Use minimum water for all the domestic purposes.
- Use drip irrigation to improve the irrigation efficiency and reduces evaporation.
- Reuse the soapy water, after washing clothes, for washing off the country yards, drive ways, etc.
- Build rainwater harvesting system in our house.
- The wasted water, coming out from kitchen, bath tub, can be used for watering the plants.

3. Conservation of Soil

- While constructing the house don't uproot the trees as for as possible.
- Grow different types of plants, herbs, trees and grass in our garden and open areas, which bind the soil and prevent its erosion.
- Soil erosion can be prevented by the use of sprinkling irrigation.
- Don't irrigate the plants using a strong flow of water, as it will wash off the top soil.
- Use mixed cropping, so that some specific soil nutrients will not get depleted.
- Use green manures in the garden, which will protect the soil.

4. Conservation of Food Resources

- Eat only minimum amount of food. Avoid over eating.
- Doesn't waste the food instead give it to someone before getting spoiled.
- Look only required amount of the food.
- Don't cook food unnecessarily.
- Don't store large amounts of food grains and protect them from damaging insects.

5. Conservation of Forest

- Use non-timer products.
- Plant more trees and protect them.
- Grassing, fishing must be controlled.
- Minimize the use of papers and fuel wood.

- Avoid of executing developmental work like dam, road, construction in forest areas.

3.6.5 Equitable Use of Resources for Sustainable Lifestyles

Scarcity of the resources is the burning problem of modern technology. The twenty first century will see growing human needs for resources as many parts of the world are using natural resources at a rate faster than the natural processes can recharge it.

Natural resources are limited. For example, an existing water sources are being subjected to heavy pollution. Global climatic changes are changing the quality of fresh water sources as a consequence of the unknown effects on the hydrological cycle.

Sustainable development is currently being a field of development, planning and other associated aspects. In the light of the self-defeating current mode of development and recurrent natural calamities, people are urged to ponder over the faults, lacunae, shortcomings, discrepancies and limitations of the ongoing process and production system.

It is an essential to sustain the natural resources. We should conserve the natural resources so that it may yield sustainable benefit to the present generation while maintaining its potential to meet the needs of the future generation. There are three specific objectives to conserve living resources:

- To preserve the biodiversity.
- To ensure that any utilisation of the ecosystem is sustainable.
- To maintain the essential ecological processes.

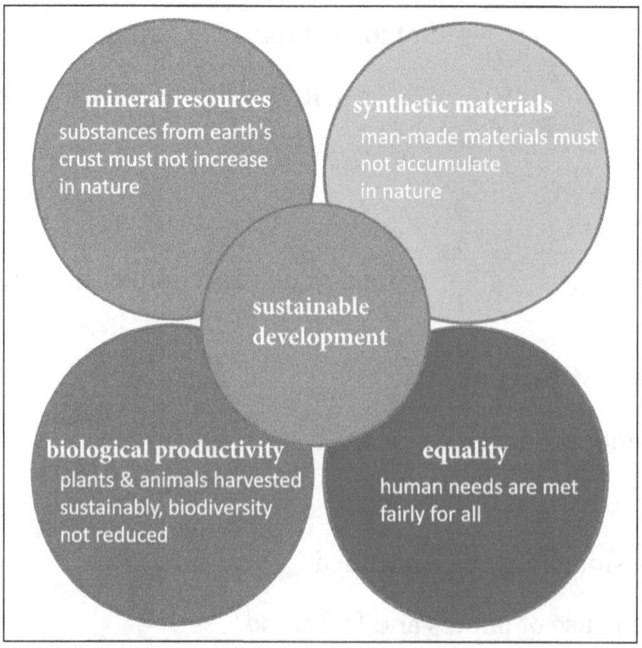

Resource management should be less energy-intensive, less cost-intensive and more viable in terms of economy, ecology and culture, suitable to local ecology and needs of the people. The Sri Lankan team, for example studied traditional paddy irrigation systems as a model for water management.

Its report note that from the 5th century B.C. through the 12th century A.D., Sri Lanka developed a technologically advanced civilization based on the intricate system of rainwater conservation and irrigation. Water users were collectively and individually responsible for maintenance of the irrigation systems and customary laws, known as Sirit, were established governing the water use and the related aspects of life.

Similarly, the italics system is a system of farmer-managed canal irrigation, which has been in operation for more than 300 years in Dhule and Nashik district of northwestern Maharashtra.

3.7 Introduction to Environmental Biochemistry: Proteins

Biochemistry

The focal point of biochemistry and biochemical aspects of toxicants is the cell, the basic building block of living systems where most life processes are carried out.' Bacteria, yeasts and some algae consist of single cells. However, most living things are made up of many cells. In a more complicated organism the cells have different functions. Muscle cells, liver cells, skin cells and brain cells in the human body are quite different from each other and do different things. Cells are divided into two major categories depending upon whether they have a nucleus. Eukaryotic cells have a nucleus and prokaryotic cells do not. Prokaryotic cells are found predominantly in the single-celled organisms such as bacteria. Eukarymic cells occur in the multicellular plants and animals - higher life-forms.

Proteins

Proteins are nitrogen-containing organic compounds that are the basic units of living systems. Cytoplasm. The jelly-like liquid filling the interior of cells, consists largely of protein. Enzymes, which act as catalysts of life reactions, are proteins. Proteins are made up of airline acids joined together in huge chains. Amino acids are organic compounds that contain the carboxylic acid group, $-CO_2H$ and the ammo group, $-NH_2$. They are sort of a hybrid of carboxylic acids and amines.

Proteins are polymers or macromolecules of amino acids containing from approximately 40 to several thousand amino acid groups joined by the peptide linkages. Smaller-molecule amino acid polymer, contains only about 10 to about 40 amino acids per molecule, are termed as polypeptides. A portion of the amino acid left after the elimination of Hp during polymerization is called a residue. The amino acid sequence of these residual is designated by a series of three-Inter abbreviations for the amino acid.

Natural amino acids all have the following chemical group:

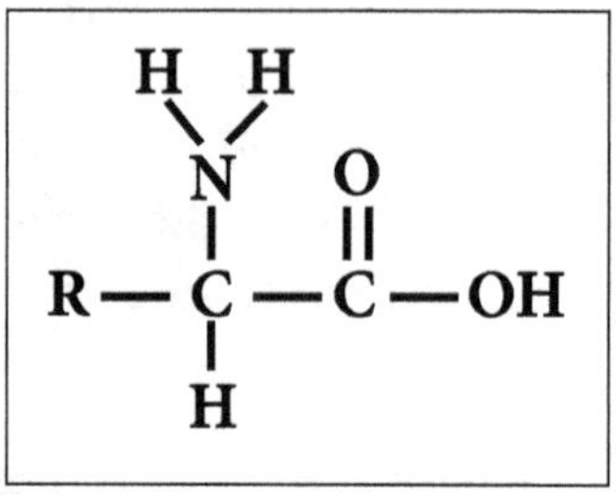

In this structure, the —NH2 group is always bonded to the carbon next to the $—CO_2H$ group. This is termed as the alpha location. So natural amino acids are termed as alpha-amino acids. Other groups, designated as R, are attached to the basic alpha-amino acid structure. The R groups may be as simple as an atom of II found in glycine:

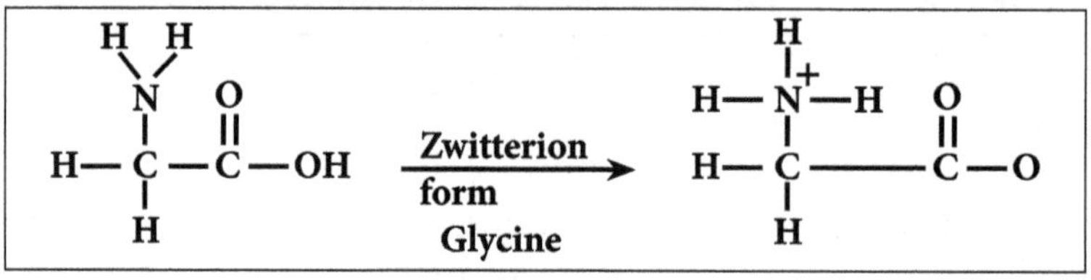

Or as complicated as the structure:

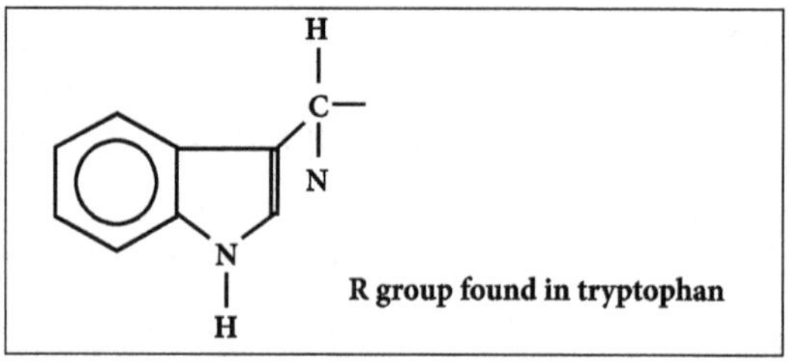

There are 20 common amino acids in proteins, example of which are shown in Figure. The amino acids are shown with uncharged $—NH_2$ and $—CO_2H$ groups. Actually, these functional groups exist in the charged zwitterion form, as shown for glycine, above. Amino acids in proteins are joined together by a specific bond called the peptide linkage. The formation of peptide linkages is a condensation process involving the loss of water.

Consider as an example, the condensation of alanine, leueine and tyrosine as shown in Figure. When these three amino acids join together, two water molecules are eliminated. The product is a tripeptide, since there are three amino acids involved. The amino acids in proteins are linked as shown for this tripeptide, except that many more monomeric amino acid groups are involved. Proteins may be divided into several major types that have widely varying functions. These are given in Table.

3.7.1 Biochemical Degradation of Pollutants

The process of bio-degradation can be defined as decomposition or destruction of pollutant molecules by the action of enzymatic machinery of a biological system. That means hardly any complex organic molecule present in the environment or any chemical which man's ingenuity and there synthetic skill may synthesize which is decomposed by this excellent and versatile machinery of enzymatic degradation.

Agents of Bio-Degradation

Though bio-degradation can be brought about by most of the living organisms, tiny microbes such as bacteria, algae, fungi, small animals etc. plays a vital role in the decomposition of various pollutants in an ecosystem. This is due to their catabolic versatility, the diversity in species composition and their ability to perform reactions at a much faster rate per unit weight as compared to other organisms. These microbes perform specific biochemical reactions which degrade macro-molecules into small units in step-wise fashion. Usually a series of organisms are required to bring the complete degradation of a large molecule or a group of molecules.

Necessary Conditions for Bio-Degradation

Though there may be some microbes which perform a number of metabolic reactions, in most of the cases only specific type of microbes are required to perform specific reactions. In many cases a number of species have to act one after the other to bring about the complete degradation.

The presence of microbes which specifically perform the task of decomposing specific structures in the pollutant molecule is, therefore, one of the basic necessities of bio-degradation. Another necessity of the biotic decomposition reaction is that contact has to be established between the microbes and the pollutant molecules which should be in the form suitable for the reaction.

It is only as a sequel to the establishment of this contact that the enzymatic machinery required for the purpose is usually brought into existence or in other words induced into being. Pollutants present in such a state which is unable to induce this machinery escape the process of bio- degradation. Availability of favourable conditions of environment such as suitable pH, water, temperature etc. in which the microbes may proliferate and the enzyme systems may operate, further speeds up the process of bio-degradation.

Nature of Bio-Degradation Reaction

Bio-degradation reactions break large molecules into smaller units so that they are rendered as harmless and are eliminated from the system. These reactions may include oxidation, reduction, hydrolysis and different types of structural re-arrangements. Simple carbohydrates, proteins, fats, nucleic acids etc. are quickly degraded.

Molecules having highly branched chains, simple or fused aromatic rings, an increased degree of substitution, cyclo-paraffins etc. are decomposed with considerable difficulty.

Degradation of these molecules involves step-wise removal of substituted groups of atoms, destruction of the ring structures and progressive alteration of constituent units etc. This explains the persistence of the petroleum crude, poisonous pesticides and other products, plastics and plasticizers etc.

Generally most of the lipophilic compounds which pose a greater hazard to a biological system as they are capable of diffusing through biological membranes, are converted to water soluble or hydrophilic ones. This results in the restriction of their movement within the biological system and in due course of time, they are excreted. To bring about this conversion, synthetic reactions are also employed by the biological system. Conjugation with glucuronic acid, sulphates, amino-acids, acetylation and methylation are some of the common reactions which transform a pollutant molecule into water soluble compound. Once transformed, the pollutant is excluded from the biological system.

Unique Dynamism of Bio-Degradation Reaction

The process of bio-degradation has a remarkable capability to adjust its activity according to the amount of pollutant present in the system. The rate of bio-degradation depends upon the ratio between the concentration required to saturate the system and the concentration of pollutant within the biological system. Below the saturation, degradation reactions decomposes a constant fraction of the total content of pollutant per unit time. Therefore, if a concentration of 200mg of a pollutant is reduced to half of its original value in one hour, implying a rate of 100mg per hour, a concentration of 20 mg shall take the same time to get reduced to 10mg which however shall occur at a rate of 10 mg per hour.

Larger loads of pollutants, which pose a greater toxicological risk are decomposed and faster rate which slows down as the concentration of the pollutants within the system diminishes. It is for this reason that the term half-life is used to measure the period of retention of pollutants or any foreign substance within a biological system. Half-life of a foreign substance in a biological system denotes the duration of time within which its concentration is reduced to half of its original value.

Small quantities of pollutants regularly entering a biological system are efficiently decomposed and eliminated. Larger amounts which the system is unable to decompose completely before the entry of fresh dose tend to accumulate within the system. As the concentration of foreign substance raises the rate of degradation or elimination is also speed up. This results in a state of dynamic equilibrium between the intake and rate of decomposition which adjusts itself in such a way as to keep the concentration of the pollutant at a reasonably constant level.

3.7.2 Bioconversion of Pollutants

Bioconversion plants generate more air pollution per unit of energy out-put than any other fuel production process except possibly cellulose hydrolysis processes. In bioconversion, unlike other processes, the production rate of air pollutants is due more to the energy material used for process heat than to the feedstock or the process technology used.

If coal is used to fire the boilers to create heat, then large quantities of nitrogen oxides (NOx), sulfur oxides (SOx), particulates and volatile organic compounds are emitted. If petroleum is used for process heat, similar pollutants, but in smaller quantities, are generated. If cellulosic biomass is used, then particulates are the most significant pollution generated. Those estimates are based on engineering studies conducted by four different research laboratories and design firms for eight hypothetical plants ranging in site from 1 to 50 million gallons per year.

3.7.3 Field Study of Local Area to Document Environmental Assets: River , Forest , Grassland , Hill and Mountain

To document, the environmental assets of an area, a detailed analysis of the habitat factors like topography, soil condition and climate must be made in order to reach an understanding of the environment of the area.

The floristic composition should also be carefully observed. It is also essential to consult a good topographical and geological maps and to consult a local meteorological station to understand the climate. It is advantageous for a beginner to concentrate the study of environmental assets of a selected sites.

- River: Rivers are inland flowing fresh water bodies. They differ widely in physical and chemical conditions, oxygen content, temperature, speed and volume of water. Gullies and ravines are common on the banks of rivers due to the erosion of soil by run off water. during the rainy season. Soil on the bank is poor in nutrients due to leaching and remains dry during the winter and summer season. Thorny bushes are the dominant plants on the banks however a number of annual herbs and grasses appear during the rainy season.

- Forest: A forest is a biotic community which is predominantly composed of trees shrubs or any woody vegetation and often with a closed canopy (overhanging covering). It differs very much in composition and density from meadows and pastures. The soil is moist and rich in organic matter. The air is cool and humid due to transpiration of the plants.

- Grasslands: A grassland has treeless herbaceous plant cover, dominated by a wide variety of grass species. Associated with grasses are several herbs especially legumes. The most remarkable aspect of grassland is its extensive root system highly ramifying through the soil horizons. The sod is rich in nutrients.

- Hill Mountains: The soil on a hill mountain site is rocky with large quantities of pebbles and silt. The soil is unstable due to continuous erosion on the slopes. The moisture content of the soil is low due to rapid percolation of water how the slope. As one ascends a mountain moisture content of the air increases and also the wind velocity. Because of the interaction of various climatic factors, mountains show different vegetation zones at different heights. The major zones are tropical, subtropical, temperate and alpine. In western Himalayas the tropical and subtropical vegetation consists of mixed deciduous forest, which is replaced at places by Savanna and dry thorn scrub. The temperate vegetation is made up of mixed coniferous forest having both conifers and broad leaved trees. The alpine zone has small shrubs and some grasses due to high wind velocity, shorter growing season, winter drying and higher light intensity during the day.

The observations related to the visit 'flocs' area may be presented in the following way:

S. No.	Environmental characters	Observations
1.	Climate	
2.	Topography	
3.	Plants	
4.	Soil conditions	
5.	Animals	
6.	Impact of degradation(if any)	
7.	Degradation(if any)	

Chapter 4

Social Issues and the Environment

4.1 From Unsustainable to Sustainable Development

To build up the sustainable development, the following approaches or methods are proposed:

Developing Appropriate Technology

It is the one which is locally adaptable. Ecofriendly, resource - efficient and culturally suitable. It uses local labors, less resources and produce minimum waste.

Reduce, Reuse and Recycle (3-R) Approach

It insists on optimum use of natural resources to use it again and again instead of throwing it on the waste land or water and recycling the material into further products. It reduces pressure on our natural resources and reduces waste generation and pollution. Providing environmental education and Awareness: By providing environmental education and awareness, the thinking and attitude of people towards our earth and the environment can be changed.

Consumption of Renewable Resources

In order to attain sustainability, it is very important to consume the natural resources in such a way that the consumption should not exceed regeneration capacity.

Conservation of Non-Renewable Resources

Non-renewable resources should be conserved by recycling and reusing.

Population Control

By controlling population growth, we can make sustained development.

4.2 Urban Problems Related to Energy

Urban center use enormous quantities of energy. In the past, urban housing required relatively smaller amounts of energy than we use at present. Traditional housing in India had required very little temperature adjustments as the material used, such as wood and bricks, handled temperature changes better than the current concrete, glass and steel of ultramodern building.

Cities are the main centers of economic growth, trade, education, innovations and employment. Until recently a large majority of human population lived in rural areas and their economic activities cantered on agriculture, fishing, cattle, rearing, hunting or some cottage industry.

It was about two hundred years ago with the dawn of industrial are the cities showed rapid development. Now about 50% of world population lives in the urban areas and there is increasing movement of rural folk to cities in search of employment.

The urban growth is very fast that it is becoming difficult to accommodate all the industrial, commercial and residential facilities within a limited municipal boundary. As a result there is spreading of cities into the suburban or rural areas too, this phenomenon is referred as “urban sprawl”.

In developing countries too, urban growth is very fast and in most of the cases, it is uncontrollable and unplanned growth. In contrast to the rural set up, Urban set up is densely populated, consumes a lot of energy and materials and generates a lot of waste.

Energy use is closely related to development in industry, communication, transport, commercial, household and agricultural activities. Energy requirement of urban population is much higher than that of rural ones. This is because urban people have a higher standard of life and their lifestyle demands more energy inputs in every sphere of life.

In urban areas, the need of energy is increasing by the leaps and bounds. Moreover, countries use energy in an uneven manner in the world. In developed countries, the amount of energy used is much more compared to developing countries.

4.3 Water Conservation

The following strategies can be adopted for conservation of water:

Reducing Evaporation Losses

Evaporation of water in humid region can be reduced by placing horizontal barriers of asphalt below the soil surface, which increases the water availability and crop yield.

Reducing Irrigation Losses

The water losses during irrigation can be reduced by the following methods:

- Sprinkling irrigation and drip irrigation conserves water by 30-40%.
- Growing hybrid crop varieties, which require less water, also conserve water.
- Irrigation in early morning or later evening, reduces evaporation losses.

Reuse of Water

- Treated wastes can also be used for fertile irrigation.
- Grey water from washing's, bathrooms, etc., may be used for wasting cars, watering gardens.

Preventing Wastage of Water

Wastage of water can be prevented by:

- Closing the taps when not in use.
- Repairing any leakage from pipes.
- Using small capacity of taps.

Decreasing Run-off Losses

Run off on most of the soils can be reduced by allowing most of the water to infiltrate into the soil. This can be done by using contour cultivation or terrace farming.

Avoid Discharge of Sewage

The discharge of sewage into natural water resources should be prevented as much as possible.

4.3.1 Rain Water Harvesting

Rainwater harvesting is the accumulation and deposition of rainwater for reuse on site, rather than allowing it to run off. Its uses include water for garden, water for livestock, indoor heating, water for irrigation, water for domestic use with proper treatment and for houses etc.

In many places the water collected is just redirected to a deep pit with percolation. The harvested water can be used as drinking water as well as for storage and other purpose like irrigation.

Objectives of rain water harvesting:

- To meet the increasing demands.
- Raise the water table by recharging the ground water.
- Reduce ground water contamination.
- Reduce the surface run off loss & soil erosion.
- Increase in hydro static pressure.
- Minimize water crisis & water conflicts.

The following figure shows the working system:

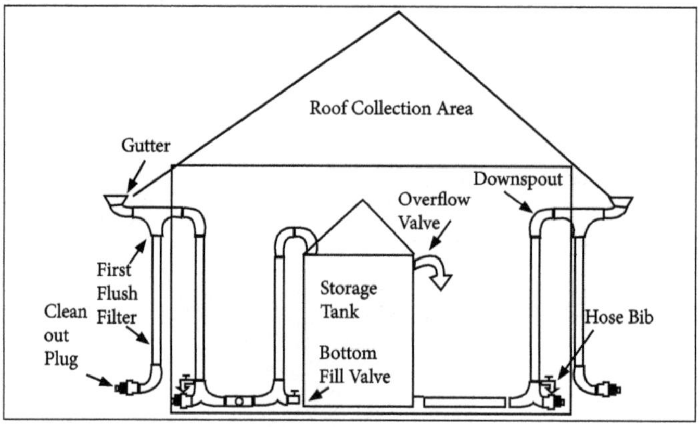

Rain water harvesting.

The advantages of rain water harvesting are:

- Reduction in the use of current for pumping water.
- Migrating the effects of droughts and achieving drought proofing.
- Increasing the availability of water from well.
- Rise in ground water level.
- Minimizing the soil erosion and flood hazards.
- Upgrading the social and environmental status.
- Future generation assured of water.

4.3.2 Waste Shed Management

The source of the waste material are agriculture, mining, industrial and municipal wastes.

Objectives:

- To minimize the risk of floods.
- For improving the economy.
- For developmental activities.
- To generate huge employment opportunities.
- To promote forestry.
- To protect soil from erosion.

Water shed management Techniques:

- Trenches (Pits).
- Dam.
- Farm pond.
- Underground barriers.

Maintenance of Watershed:

- Afforestation.
- Water harvesting.
- Scientific mining & Quarrying.
- Reducing soil erosion.
- Minimizing livestock population.
- Public participation.

4.4 Resettlement and Rehabilitation of People: Problems and Concerns

Rehabilitation of People

People are forced to evacuate from their land due to both natural and manmade disasters. Natural disasters like earthquakes, cyclones, tsunami etc.

Render thousands of people homeless and sometime even force them to move and resettle in other areas. Similarly, developmental projects like construction of dams, roads, flyovers and canals displace people from their home. Thus, resettlement refers to the process of settling again in a new area. Rehabilitation means restoration to the former state.

Resettlement Issues

As per the World Bank estimates, nearly 10 lakhs people are displaced worldwide for a variety of reasons.

Meager Compensation

The compensation for the land lost is often not paid, it is delayed or even if paid, is too small for both in monetary terms and social changes forced on them by these big developmental projects.

Little or No Support

Displacement mainly hits tribal and rural people who usually do not figure in the priority list of political authorities or parties. Why do we think that the rural people have become the red Robin hoods of today which are the Maoists.

Lack of Facilities

When people are resettled in a new area, basic infrastructure and amenities are not provided in that area. Very often, temporary camps become permanent settlements. It is also an important problem of displacement or resettlement that people have to face.

Loss of Livelihood

Displacement is not a simple incident in the lives of the displaced people. They should leave their ancestral land and forests on which they are depended for their livelihood. Many of them does not posses any skills to take up another activity or pick up any other occupation. Usually, the new land that is offered to them is of poor quality and the refugees are unable to make a living.

Increase in Health Problems

Lack of nutrition due to the loss of agriculture and forest based livelihood, lead to the general decline in the health of the people. People are used to traditional home remedies. But the herbal remedies and plants gets submerged due to the developmental projects.

Increase in Stress

Resettlement disrupts the entire life of the people. They are unable to bear the shocks of purposelessness and emptiness created in their life. The payment of compensation to the head of the family often lead to bitter quarrels over sharing of compensation amount within the family, leading to stress and even withering of family life. Moreover, the land ownership has a certain prestige attached to it that cannot be compensated for, even after providing the new land. With the loss of property and prestige, the marriages of young people also become difficult as people from the outside villages are not willing to marry their daughters to the refugees.

Secondary Displacement

The occupational groups which are residing outside the submergence area but depending on the area for livelihood also experience unemployment. Laborers, village artisans, petty traders etc, lose their living.

Loss of Identity

Tribal life is based on the community. The tribal are very simple people who have a lifestyle of their own.

Displacement have a negative impact on their livelihood, cultural and spiritual existence in the following ways:

- Inter community marriages, folk songs, cultural functions and dances does not take place among the displaced people. When they are resettled, it is usually individual based resettlement, which ignores communal character.
- Break up of the families and communities are the important social issues of displacement. The women suffer the most as they are deprived even a little compensation.
- Resettlement increases the poverty of tribal due to the loss of land, livelihood, food insecurity, jobs, skills etc.
- Loss of the identity of individuals and loss of connection between the people and the environment is greatest loss in the process. The indigenous knowledge that they have regarding the herbal plants and wildlife are lost.
- The laws of land acquisition does not pay attention to the idea of communal ownership of property which increases stress within the family.
- The tribal people are not familiar with market trends, prices of commodities and policies. As such, they are exploited and get alienated in modem era.

My mother often says about the plight of people from Nepal who came newly to city of Gangtok. Earlier, when they demanded four meters cloth piece, the shopkeeper used to measure the same from all the four sides of the piece. Actually it was only one meter in length, but they used to charge the price of four meter cloth.

Objectives of Rehabilitation

The following objectives of the rehabilitation should be kept in mind before people are given an alternative site for living:

- Tribal people should be allowed to live along the lives of their own patterns and others should avoid imposing anything on them.
- They should be provided means to develop their own traditional art and culture in possible way.
- Villagers should be given the option of shifting out with the others to enable them to live community based life.
- The people displaced should get an appropriate share in the fruits of the development.
- If resettlement is not possible in the neighbor area, priority should be given to development of the irrigation facilities and supply of basic inputs for agriculture, drinking water, wells, grazing ground for the schools for the children, cattle, primary health care units and other amenities.
- Removal of poverty should be one of the objectives of rehabilitation.
- The elderly people of the village should be involved in the decision making.
- The displaced people should be given employment opportunities.
- Resettlement should be in the neighborhood of their own environment.
- Villagers should be taken into confidence at every stage of implementation of the displacement and they should be educated, through public meetings, discussion about the legalities of the Land Acquisition act and other rehabilitation provisions.

4.5 Role of Non-Governmental Organization

Non-government organization is a people's organization. It could be a real force for the protection of the environment in any community. The success of India's environmental Program depends mainly on the awareness and consciousness of the people. The following are some environmental organization in India:

- Voluntary organizations help the government in local environmental and interact at grass root level.
- The NGO, 'Mahila Mandal's creates awareness amongst the women of remote villages to educate them and help them to become healthy and help them to become healthy economically self-dependent.
- Narmada Bachao Andolan was organized by Kalpav Viksh.

- Kerala Sastra Sahitya Prashad, Centre for Science and Environment (CSE). The report as per CSE on permissible limits of pesticides in the Cola drinks sensitized the public all over the country.
- Bombay Natural History Society (BNHS), Madras University society(NMS), Chennai world wide fund for nature (NWF). India is some of NGGs taking many steps for the environmental protection and health.
- Chipko movement works for the conservation of trees by Darhote Gram Swarajya Mandal in Gopeshwar.
- NGO's also acts as action group.

4.6 Environmental Ethics: Issues and Possible Solutions

Environmental ethics refers to the issues, principles and guidelines relating to human interactions with their environment. It also means that effort must be taken to protect an environment and to maintain its stability from the hazardous chemical pollutants.

Functions of Environment

- It is the life supporting medium for all organisms.
- If provides food, air, water and other important natural resources to the human beings.
- It disintegrates all the waste materials discharges by the modern society.
- It moderates the climatic conditions of the soil.
- A healthy economy depends on a healthy environment.

Environmental Problems

- Deforestation activities.
- Population growth and urbanization.
- Pollution due to discharge of effluent and smoke discharge from the industries.
- Water scarcity.
- Land degradation and degradation of soil fertility.

Solutions to Environmental Problems

The environmental can be protected due to the following activities:

- Reduce the waste of matter and energy resources.

- Recycle and reuse as many of our waste products and resources as possible.
- Over exploitation of natural resources must be reduced.
- Soil degradation must be minimized.
- Sustainable development is essential on, conservation of resources, harvesting of non-conventional energy and waste management.
- Biodiversity of the earth must be protected.
- Reduce population and increase the economic growth of our country.

4.7 12 Principles of Green Chemistry

The following list outlines an early conception of what would make a greener chemical, process or product.

Prevention

It is better to prevent the waste than to treat or clean up waste after it has been created.

Atom Economy

Synthetic methods are designed to maximize the incorporation of all the materials used in the process into the final product.

Less Hazardous Chemical Syntheses

Wherever practicable, synthetic methods should be designed to use and generate the substances that possess very little or no toxicity to human health and the environment.

Designing Safer Chemicals

Chemical products should be designed in such a way to affect their desired function while minimizing their toxicity.

Safer Solvents and Auxiliaries

The use of auxiliary substances should be made unnecessary wherever possible and innocuous when used.

Design for Energy Efficiency

Energy requirements of chemical processes should be recognized for their environmental and economic impacts and should be minimized. If possible, synthetic methods should be conducted at the ambient temperature and pressure.

Use of Renewable Feed Stocks

A raw material or feed stock should be renewable rather than depleting whenever they are technically and economically practicable.

Reduce Derivatives

Unnecessary derivation should be minimized or avoided if possible, because such steps require additional reagents and can generate waste.

Catalysis

Catalytic reagents are superior to the stoichiometric reagents.

Design for Degradation

Chemical products should be designed so that at the end of their function they break down into innocuous degradation products and do not persist in the environment.

Real-Time Analysis for Pollution Prevention

Analytic methodologies need to be further developed to allow for real-time, in-process monitoring and control prior to the formation of hazardous substances.

Inherently Safer Chemistry for Accident Prevention

Substances and the form of substance used in a chemical process should be chosen to minimize the potential for chemical accidents including releases, explosions and fires.

4.8 Nuclear Accidents and Holocaust

Nuclear Accidents

- Radiation may break chemical bands such as DNA in cells. This effect may be instantaneous, prolonged or delayed types. It may be even carried to future generation.
- Exposure at low dose of radiation, people do not die, but begin to suffer from fatigue, vomiting and loss of hair.
- Exposure at higher dose of radiation (400-500 rads) affects bone marrow, blood cells, natural resistance and blood to fail clot.
- Exposure at very high dose of radiation (10,000 rads) kills the organisms by damaging the tissues of heart, brain.

Nuclear accidents.

Nuclear Holocaust

It means destruction of biodiversity by nuclear equipment's and nuclear bombs. In a holocaust, a large number of human lives are totally destroyed.

Effect of Nuclear Holocaust

Nuclear Winter

Nuclear bombardment will cause combustion of wood, plastics, petroleum, forest etc. Large quantity of black soot will be carried to the stratosphere. Black soot will absorb all the UV- radiations and will not allow the radiation to reach the earth. Therefore, cooling will result. Due to this cooling effect, water evaporation will also reduce. In stratosphere there won't be significant moisture to rain-out the thick soot. Thus, due to nuclear explosions, a process known as opposite to global warming will occur. This is called nuclear winter.

Effect of Nuclear Winter:

- Lowers the global temperatures, even in summer the temperature will be at around freezing temperature.
- Crop productivity will be reduced causing famines and human sufferings.

It ignites all combustible material, destroys all the living beings and materials:

Examples of Nuclear holocausts:

- Nuclear war: Japan, Hiroshima and Nagasaki are the examples of nuclear holocaust, which had happened during world war II.

- At Chernobyl: When the operators lost the control of a water- cooled, graphite moderated reactor during a low power tests at Chernobyl in Ukraine, the nuclear reactor exploded.

4.8.1 Case Studies

Nuclear Holocaust in Japan

In 1945, two nuclear bombs were dropped on Hiroshima and Nagasaki cities in Japan. This holocaust killed about 100,000 people and totally destroyed the city. This explosion emitted forceful neutrons and gamma radiations. The radioactive strontium (Sr90*) liberated in the explosion resembles calcium and has the property of replacing calcium in the bones. As a result large scale bone deformities occurred in the inhabitants of these cities.

Chernobyl Nuclear Disaster

On April 26th 1986, the world worst nuclear explosion occurred in one of the reactors in Chernobyl power plant in Ukraine. This disaster was caused mainly due to poor reactor design and human error.

Impacts

Chernobyl accident killed more than 80,000 people and 3.5 million people have fallen ill because of the accident. Nearly 1, 35,000 people have to be evacuated at around Chernobyl.

Chernobyl victims includes:

- 31 deaths at the time.
- An estimate 2000 extra cases of career in Europe over the next 50 years.
- A much greater loss of life and damage to health can still be attributed to the real industry.

Control Measures

- Suitable precautions are to be taken and training must be given to people for handling these materials to avoid accident.
- Constant monitoring of the radiation level has to be carryout out, limit exposure to the workers.
- Regular checks and control measures are done by Atomic Energy Regulatory Board under the Department of Atomic Energy.

Bhopal Gas Tragedy

The world's worst industrial accident occurred in Bhopal city, M.P. on the night of 3rd December 1984. It happened at Union Carbide India Ltd., who manufactures some carbonate pesticides, using methyl isocyanate (MIC).

The reactor exploded, due to failure of its cooling system and 40 tons of MIC leaked into the atmosphere.

Effects

- Methyl-iso-cyanate is a toxic gas, lower concentrations of which affects lungs and lyes and causes irritation in the skin.
- Higher amounts of MIC, remove oxygen from the lungs and can cause death.

Effect in Bhopal

The gas (MIC) spread over 40 sq.km. area. About 5000 persons were killed and 65,000 people suffered from severe eye, respiratory, neuro muscular, gastrointestinal disorders. Nearly 1000 persons become blind.

4.9 Wasteland Reclamation

Waste Land

The land which is not in use is called waste land. The waste land is unproductive, unfit for cultivation, graying and other economic uses. About 20% of the geographical area of India is waste land.

Types of Waste Lands

Wastes lands can be divided into two types:

Uncultivated Waste Lands: These lands cannot be brought under cultivation.

- Examples: Barren rocky areas, hilly slopes, stony or leached or gully land or sandy deserts, etc.

Cultivable Waste Lands: These are cultivable but not cultivated for more than five year. Cultivable waste lands are important for agricultural purposes.

- Examples: Degraded forest lands, gullied water, logged and marsh lands, saline lands, etc.

Causes of Waste Land Formation

- Due to soil erosion, deforestation, over graying, water logging, salinity.
- The increasing demand for firewood and exercise use of pesticides.
- Over exploitation of natural resources.
- By the sewage and industrial wastes.
- Mining activities destroy the forest and cultivable land.

Objectives or Need of Waste Land Reclamation

- To improve the physical structure and quality of the soil.
- To prevent soil erosion, flooding and landslides.
- To avoid over exploitation of natural resources.
- To improve the availability of good quality of water for agricultural purposes and industrial operations.
- To conserve the biological resources and natural ecosystem.
- To provide a source of income to the poor.

Methods of Waste Land Reclamation

- Drainage: Excess water is removed by artificial drainage. This process is used for water logged coil reclamation.
- Leaching: Leaching is the process of removal of salt from the salt affected soil by applying access amount of water. Leaching is done by dividing the field in small plots. In continuous leaching 0.5cm to 1 .0cm water is required to remove 90% of soluble salts.
- Irrigation practice: High frequency irrigation with controlled amount of walls help to maintain better water availability in the Land.
- Green Kamures and Bio sett-lingers: Application of green Kamure is found to improve the saline soils.
- Application of Gypsum: Soil sodicity can be reduced with gypsum. Calcium of gypsum replaced sodium from the unchangeable sites. This process converts clay back into calculus clay.
- Social Forestry Programme: These programmers induce string plantation on road, land-sides, degraded forest land, etc.

4.9.1 Consumerism and Waste Products

Consumerism refers, to the consumption of resources by the people. It is an organized movement of citizens and government. The special concentration is given to improve the rights and power of the buyers in relation to the sellers.

Consumerism is related to both increase in population as well as increase in our demand due to charge in lifestyle. In the modern society, our needs have a increased and so consumerism of resources has also increased.

Traditionally favorable rights of sellers:

- The right to introduce any product.
- The right to charge any price.
- The right to spend any amount to promote their product.
- The right to use incentives to promote their products.

Traditional buyer rights:

- The right to buy or not to buy.
- The right to expect a product to be safe.
- The right to expect the product to perform as claimed.

Important information's to be known by buyers:

- Ingredients of a product.
- Manufacturing date and expiry date.
- Whether the product has been manufactured against as established law of nature or involved in rights violation.

Objectives of consumerism:

- It improves the rights and power of the buyers.
- It involves making the manufacturer liable for the entire life cycle of a product.
- It forces the manufacturer to reuse and recycle the product after usage.
- The items which are very difficult to decompose like polymeric goods, computers, televisions etc., can be returned to manufacturer for reclaiming useful parts and disposing the rest.
- The reusable packing materials like bottles can be taken back to the manufacturer. It makes the products cheaper and avoids littering and pollution's.
- Active consumerism improves human health and happiness and also it saves resources.

Sources of Wastes: The sources of the waste materials are agriculture, mining, industrial and municipal wastes.

Examples for waste products: It includes glass, papers, garbage, plastics, soft drink canes, metals, food wastes, automobile wastes, dead animals, construction and factory wastes.

E-Waste (Electronic and Electric Wastes)

- Electronic equipment's like computers, printers, mobile, phones, xerox machines, calculators, etc.
- After using these Instruments, they are thrown as waste.

Effects of Wastes:

- The waste released from chemical industries and from explosives is dangerous to human life.
- The dumped wastes degrade soil and make unfit for irrigation.
- E-waste contains more than 1000 chemicals, which are toxic and cause environmental pollution. In computer, load is present in monitors, cadmium in chips and cathode ray tube, PVC in cables. All these cause cancer and other respiratory problems if inhaled for long periods.
- Plastics are difficult to recycle or incinerate safely because they are non-biodegradable and their combustion produces several toxic gases.

Factors affecting consumerism and generation of wastes:

- People over Population: It occurs when there are more people than the available supply of food and water. Over population causes degradation of resources, poverty and premature death. This situation occurs in less developed countries (LDC's). Thus in LDC's per capita consumption of resources and waste generation are less.
- Consumption Over-Population: It occurs when there are less people than the available resources. Due to luxurious life-style per capita consumption of resources is very high. If the consumption is more, the generation of waste is also more and greater is the degradation of environment.

4.10 Environment Protection Act

This is a general legislation law in order to rectify the gaps and laps in the above Acts. This Acts empowers the central government to fix the standards for quality of air water, soil and noise and to formulate procedures and safe guards for handling of hazard substance.

Objective of environmental act:

- To protect and improve the environment.
- To prevent hazard to all living creatures and property.
- To maintain harmonious relationship between humans and their environment.

Important features of environment act:

- The Act further empowers the Government to lay down procedures and safe guards for the prevention of accidents which cause pollution and remedial measures if an accident occurs.
- The Government has the authority to close (or) prohibit (or) regulate any industry (or) its operation, if the violation for the provisions of the Act occurs.
- If the violation continuous, an additional fine of Rupees five thousands per day may be improved for the entire period of violation of rules.
- The Act fixes the liability of the offense punishable under Act on the person who is directly in change.

4.10.1 Air Act

Air (Prevention and Control of Pollution) Act, 1981

This Act was enacted in the conference held at Stockholm in 1972. It deals with the problems relating to air pollution. It envisages the establishment of central and state control boards endowed with absolute powers to monitor air Quality.

Objectives of air act:

- To prevent, control and abatement of air pollution.
- To maintain the Quality of air.
- To establish a board for the prevention and control of air pollution.

Important features of Air Act:

- The central board may lay down the standards.
- The central board co-ordinates and settle disputes between state boards, in addition to providing technical assistance.
- The state boards are empowered to lay down the standards for emissions of air pollutants from industrial units or automobiles.

- The state boards are to collect and disseminate information related to air pollution.
- The state boards are to examine the manufacturing orders.
- The state board can adhesive the state government to declare certain heavily polluted areas as pollution control areas.
- The directions of the central board are mandatory on state boards.
- The operation of an industrial unit is prohibited in a heavily polluted areas:
    - Violation of law is punishable with imprisonment for a term which may extend to three months.

4.10.2 Water Act

The Water Act was enacted by Parliament Act, 1974 for the prevention and control of water pollution and the maintaining or restoring of wholesomeness of water.

The relevant provisions of this act are given below:

- Officials of pollution control board can take samples of water effluent from any industry, stream or well or sewage sample for the purpose of analysis.
- Officials of the state boards can enter any premises for the purpose of examining any record, plant, register or any of the functions of the Board entrusted.
- No person shall discharge any poisonous, noxious or any polluting matter into any stream or well or sewer or on land:
    - Establish or take any step to establish any industry, operation or process or any treatment and disposal system for any extension or addition there to, which is likely to discharge sewage or trade effluent into a stream or well or sewer or on land.
    - Bring into use any new or altered outlet for the discharge of sewage.
    - Begin to make any new discharge of sewage.
- The state board may grant consent to the industry after satisfying itself on the pollution control measures taken by the unit or refuse such consent for reasons to be recorded in writing.
- Any person aggrieved by the order made by the State Board may within thirty days from the date, on which the order is communicated, prefer an appeal to such authority.

- The State Board can direct any person who is likely to cause the pollution of water in the street or well to desist from taking such action as is likely to cause its pollution or to remove such matters as specified by the Board through court.
- Pollution control board can issue any directions to any person, officer or authority and such person, officer or authority shall be bound to comply with such directions. The directions include the power to direct.
  - The closure, prohibition of any industry.
  - Stoppage or regulations of supply of electricity, water or any other services.
- If any who has been convicted of any offense is again found guilty of an offense involving a contravention of the same provision shall be on the second and on every subsequent conviction be punishable with imprisonment for a term which shall not be less than two years but which may extend to seven years with fine.
- Whoever contravenes any of the provisions of this act or fails to comply with any order given under this act for which no penalty has been elsewhere provided in this act, shall be punishable with imprisonment which may extend to 3 months or with fine which may extend to Rs.10, 000 or with both.

4.10.3 Wildlife Protection Act

Protective Act

This act was amended in 1983, 1986 and 1991.

This act is aimed to protect and preserve wildlife. Wild life refers to all animals and plants that are not domesticated. India has rich wildlife heritage. It has 350 species of mammals, 1200 species of birds and about 20,000 known species of insects. Some of them are listed as 'endangered species' in the wildlife (Protection) act.

Wildlife is the integral part of our ecology and plays an essential role in its functioning. The wildlife is declining due to human actions, the wildlife products skins, furs, feathers, ivory, etc., have decimated the populations of many species.

Wildlife populations are regularly monitored and management strategies formulated to protect them.

Important Features

- The Act covers the rights and non-rights of forest dwellers.

- It provides restricted grazing in sanctuaries but prohibits in national park.
- It also prohibits the collection of non-timer forest.
- The rights of forest dwellers recognized by the Forest Policy of 1988 are taken away by the Amended Wild life Act of 1991.

4.10.4 Forest Conservation Act

Conservation Act

This act provides conservation of forests and related aspects. This act also covers all type of forests including reserved forests, protected forests and any forested land. This Act is enacted in 1980. It aims at to arrest deforestation.

Important Features of Forest Act

- The reserved forests shall not be diverted or de reserved without the prior permission of the central government.
- This land that has been notified or registered or forest land may not be caused for non-forest purposes.
- Any illegal non forest activity within a forest area can be immediately stopped under Act.

Important features of Amendment Act of 1988

- Forest departments are forbidden to assign any forest land by way of lease or otherwise to any private person or non-government body for reafforestation.
- Clearance of any forest land of naturally grown trees for the purpose of reafforestation is forbidden.
- The diversion of forest land for non-forest uses is cognizable offense and any one who violates the law is punishable.

4.11 The Bio Medical Waste - Scheme of Labeling of Environmentally Friendly Products (Eco Mark)

The Bio-medical Waste (Management & Handling) Rules, 1998 and amendments thereof were notified under the Environment (Protection) Act, 1986 by the Ministry of Environment & Forest. The BMW Rules are comprehensive and set the duty of occupier/every bio-medical waste generator to segregate, transportation and store, treat and dispose these wastes as per the laid down provisions.

The State Pollution Control Board (SPCB)/ Pollution Control Committee (PCC) have been notified as the prescribed authority for implementing the provisions of these Rules in the respective State/UT. Central Pollution Control Board (CPCB) is regularly pursuing with the SPCB/PCC to get information on bio-medical waste management in the respective State/UT.

Bio-Medical Waste Management Scenario in the Country

As per the annual report information received for the year 2010 from the State Pollution.

Control Boards / Pollution Control Committees (SPCBs/PCCs) and Director General of Armed.

Forces Medical Services (DGAFMS), the bio-medical waste management status in the;

Country is given as follow:

- Number of healthcare facilities: 139594.
- Number of beds: 1420563.
- Number of Common Bio-medical Waste Treatment Facilities (CBWTFs): 188 + 17*.
- Number of healthcare facilities (HCFs) using CBWTFs: 98764.
- Number of HCFs having treatment & disposal facilities: 20228.
- Number of healthcare facilities applied for authorization: 77537.
- Number of healthcare facilities granted authorization: 70800.

Total number of on-site treatment equipment installed (at Healthcare facilities):

- Number of incinerators:
    - With Air Pollution Control Device: 419.
    - Without Air Pollution Control Device: 273.
- Number of autoclaves: 2710.
- Number of microwaves: 179.
- Number of Hydroclave: 13.
- Number of Shredders: 4250.

Total number of treatment equipments installed at CBWTFs:

- Number of autoclaves: 161.
- Number of incinerators: 177.
- Number of Hydroclave: 5.
- Number of microwaves: 10.
- Quantity of bio-medical waste generated in Tons/day: 355 (approx.).
- Number of Shredders: 170.
- Number of HCFs violated BMW Rules: 6653.
- Quantity of bio-medical waste treated in Tons /day: 302.0 (approx.).
- Number of Show-cause notices or Directions issued to defaulter HCFs: 5829.

There has been increase in number of common Bio-medical waste treatment facilities over the years and at present there are 205 CBWTFs (188 under operation + 17 under construction) so as to facilitate the proper treatment and disposal of bio-medical waste in the Country.

At present, with the advancement of medical sciences most of the hospitals/nursing homes are well equipped with latest instruments for diagnosis and treatment of various diseases. One of the most important aspect associated with hospitals is the safe management of the wastes; generated from these establishments, which contains human anatomical wastes blood, body fluid, disposable syringe, used bandages, surgical gloves, blood bags intravenous tubes etc.

The Bio-medical waste generated from various sources has become a problem and much more attention is being given worldwide to find out the solution of this problem. The main concern lies with the hospital waste generated from large hospitals or nursing homes as it may pose deleterious effects due to its hazardous nature.

Bio-medical wastes is a potent source of diseases, like Tuberculosis, AIDS, Hepatitis and other bacterial diseases causing serious threats to human health. Owing to the discussed potential threats, this waste needs prime attention for its safe and proper disposal.

4.11.1 Enforcement Machinery Involved in Environmental Legislation

The Environment (Protection) Act, 1986, has provisions for a fairly comprehensive coverage to deal with various aspects, including pollution, environmental degradation and protection of ecologically important areas. The Act provides for setting up of new institutional mechanisms where called for. It has empowered the concerned authorities to issue necessary directions, including closure of installations/activities causing pollution or environmental degradation. Stringent penalties with imprisonment and fine have also been prescribed in the Act.

However, an Act is only as good as it is implemented and there are several obstacles to its effective implementation. Apart from the administrative machinery at various levels, which needs to be geared up, a major hurdle is in terms of legal structures. More often than not, the legal proceedings are inordinately long-drawn and decisions are delayed in the process. To enable pollution control boards (PCBs) to function effectively, it is all the more necessary to strengthen/restructure PCBs and similar authorities. The manner in which the legal proceedings can be expedited (and avoided in certain cases) is also to be examined.

The Environment (Protection) Act, 1986, was promulgated under Article 253 of the constitution following the leakage of methyl isocyanate gas from the Union Carbide factory in Bhopal. Brought into force from November 19, 1986, the Act sought to deal with aspects relating to hazardous chemicals and siting of industries. It also highlighted the need for providing teeth for taking deterrent action, where warranted.

While the Environment (Protection) Act and other Acts as amended in the recent past have been provided with powerful teeth, the claws for the grip in terms of administrative and organizational structure required for the purpose arc lacking. In order to have an organic linkage among the concerned agencies at the central and state levels, creation of an all-India Environment Protection Service can be a possible means for building up cohesive enforcement machinery. The Constitution of India contains a direct commitment to environmental protection. The necessity to have umbrella legislation was also felt for taking appropriate measures on various aspects of environmental concern, including those that are not covered under the existing laws.

The Environment (Protection) Act, 1986

The Environment (Protection) Act, 1986, enables the central government to: 1. Take all necessary measures for protecting the quality of environment. 2. Co-ordinate the action of state governments, officers and other authorities under this Act or under any other law that is related to the objects of this Act.

- Take all necessary measures for protecting the quality of environment.
- Co-ordinate the action of state governments, officers and other authorities under this Act or under any other law that is related to the objects of this act.
- Plan and execute nationwide programme for prevention, control and abatement of environmental pollution.
- Lay down standards for discharge of environmental pollutants.

- Empower any person to enter, inspect, take samples and test.
- Establish or recognize environmental laboratories.
- Appoint or recognize government analysts.
- Lay down standards for the quality of environment.
- Lay down procedures and safeguards for handling hazardous substances.
- Restrict areas in which any industries, operations or processes may not be carried out or shall be carried out subject to certain safeguards.
- Lay down safeguards for prevention of accidents and take remedial measures in case of such accidents.
- Constitute an authority or authorities for exercising its powers.
- Issue directions to any person, officer or authority, including the power to direct closure, prohibition or regulation of any industry, operation or process or stoppage or regulation of supply of electricity, water or any other service.
- Require any person, officer, state government or authority to furnish any prescribed information.
- Delegate powers to any officer or state government or authority.

4.11.2 Central and State Pollution Control Boards

The Central Pollution Control Board (CPCB), statutory organization, was constituted in September, 1974 under the Water (Prevention and Control of Pollution) Act, 1974. Further, CPCB was entrusted with the powers and functions under the Air (Prevention and Control of Pollution) Act, 1981.

It serves as the field formation and also provides technical services to the Ministry of Environment and Forests of the provisions of the Environment (Protection) Act, 1986. Principal Functions of the CPCB, as spelt out in the Water (Prevention and Control of Pollution) Act, 1974 and the Air (Prevention and Control of Pollution) Act, 1981:

- To improve the quality of air and to prevent, control or abate air pollution in the country.
- To promote the cleanliness of streams and wells in different areas of the States by prevention, control and abatement of water pollution.

Air Quality Monitoring is the important part of air quality management. The National Air Monitoring Programme (NAMP) has been established with objectives to determine the present air quality status and trends and to control and regulate pollution from industries and other source to meet the air quality standards. It also provides the background air quality data needed for industrial siting and town planning.

Besides this, CPCB has an automatic monitoring station at ITO Intersection in New Delhi. At this station Resirable Suspended Particulate Matter (RSPM), Carbon Monoxide (CO), Ozone (O3), Sulphur Dioxide (SO2), Nitrogen Dioxide (NO2) and Suspended Particulate Matter (SPM) are being monitored regularly. This information on Air Quality at ITO is updated every week.

Fresh water is a finite resource essential for use in agriculture, industry, propagation of wildlife and fisheries and for human existence. India is a riverine country. It has 14 major rivers, 44 medium rivers and 55 minor rivers besides numerous lakes, ponds and wells which are used as primary source of drinking water even without treatment. Most of the rivers being fed by monsoon rains, which is limited to only 3 months of the year, run dry throughout the rest of the year often carrying wastewater discharges from industries or cities/towns endangering the quality of our scarce water resources. The parliament of India in its wisdom enacted the Water (Prevention and Control of Pollution) Act, 1974 with a view to maintaining and restoring wholesomeness of our water bodies. One of the mandates of CPCB is to collect, collate and disseminate the technical and statistical data relating to water pollution. Hence, Water Quality Monitoring (WQM) and Surveillance are of utmost importance.

State Pollution Control Board

The consent of the state pollution control Board is needed to:

- Take steps to establish any industry or any treatment and disposal system or any extension or addition there to which is likely to discharge or trade effluent into a stream or well or river or on land.
- Use any new or altered on that forth discharge of sewage.
- Begin to make any new discharge of sewage.

In the event of a violation of the conditions imposed, the state board may serve on the offender a notice imposing any such conditions as it might establishment, such outlet or discharge that is a violation of the conditions.

The Act further empowers the state board to order closure or stoppage of supply or electricity, water or any other services to the polluting unit. Non-compliance of the order may attract imprisonment for a term of one and half years to six years and fine which may extend to rupees five thousand for every day, of the default continues.

4.12 Disaster Management

There are three basic aspect of disaster management. These are:

1. Disaster Impact and Response

The disaster impact and response aspect includes the following operation Search and rescue, first aid, triage and stabilization of victims, hospital treatment, redistribution of patients to other hospitals if necessary, identification of dead, relief phase, managing humanitarian supplies, vaccination, rehabilitation, basic sanitation and personal hygiene.

2. Disaster Preparedness

The disaster preparedness is Program of long term development activities whose goals are to strengthen the overall capacity and capability of a country efficiently all types of emergency. The objective of disaster preparedness is to ensure the following tasks: Evaluate the risk of a particular region or the country as a whole to disaster organize communication, information and working systems, to ensure the availability of financial and other resources, to coordinate information with news media, to organize disaster simulation exercises to test the response of the mechanisms.

3. Disaster Mitigation

The disaster mitigation involves preventing hazards or to reduce the likely effect of emergencies. These measures include Flood mitigation works, appropriate land use planning, improved building codes, reduction of vulnerable structures and protection of vulnerable population.

Mitigation complements the disaster preparedness and disaster response activities.

4.12.1 Floods

Whenever the magnitude of water flow exceeds the carrying capacity of channel within the banks, the excess of water overflows on the surroundings causes floods.

Causes of Floods:

- Heavy rain, rainfall during cyclone causes floods.
- Sudden snow melt also raises the quantity of water in streams and causes flood.
- Sudden and excess release of compounded water behind dams.
- Human activities like construction of roads, building and parking space, that covers the earth's surface prevents infiltration.
- Clearing of forest for agriculture has also increases severity of floods.

Effects of Floods:

- Flood cause heavy suffering to people living in low lying areas because the houses and properties are washed away.
- Flood damage standing crops and livestock.

Flood Management

- Encroachment of flood ways should be banned.
- Building walls prevent spilling out. The flood water over flood plains.
- Diverting excess water through channels or canals to areas like lake, river, etc. where water is not sufficient.
- Build check-dam on small streams, move building off the flood plains.
- Restore wetlands, replace ground course.
- River-networking in the country also reduce flood.
- Flood forecast and flood warning are also given by the central water commission.
- Reduction of runoff by increasing infiltration through appropriate afforestation.

India Scenario-Floods

Next to Bangladesh, India is the most flood affected country in the world. Nearly 40 million hectares are affected by annual floods in India. 20% of which is present only in UR Next to UP Orissa, Andhra Pradesh, Bihar.

Case Studies

Flood in Bangladesh-1974

Every year in Bangladesh, large areas are submerged during the monsoon season. In 1974 flooding extended nearly one-half of the country and stagnated for more than one month.

Nearly 1500 people died in the flood and 2, 80,000 people died from subsequent disease and starvation.

4.12.2 Earthquake

An earthquake is a sudden vibration caused on the earth's surface due to the sudden release of tremendous amount of energy stored in the rocks under the earth's crust.

Occurrence: The earth's crust has several tectonic plates of solid rock. These plates move slowly along their boundaries. When friction prevents these plates from stepping, stress develops and results in sudden fractures along the fault lines within the plates. This causes earthquakes and the violent vibrations in the earth.

Causes:

- Earthquakes are caused due to dis-equilibrium in any part of the earth crust. This dis-equilibrium is caused by volcanic eruptions, hydro-static pressure of man-made water bodies like reservoirs, dams and lakes which in turn causes movements of plates.
- Underground nuclear testing.
- Decrease of underground water level.

Severity of an Earthquake

The severity of an earthquake is generally caused by its magnitude on Richter scale.

Richter Scale	Severity of Earthquake
Less than	Insignificant
4 4 - 4.9	Minor
5- 5.9	Damaging
6- 6.9	Destructive
7- 7.9	Major
More than 8	Great

- The shocks produced by earthquakes in hilly and mountainous area may cause landslides, which damage the settlements and transport systems.
- It also collapses houses and other structures due to poor construction and the people die in thousands depending on the severity of the earthquake.
- Severe earthquake results in deformation of ground surface.
- Tsunami, the seismic waves caused by earthquakes travel through sea water and generate high sea waves called tsunami. This causes great loss of life and property.

Earthquake Management

- Damage to property and life can be prevented by constructing earthquake-resistant buildings in the earthquake prone areas.
- Wooden houses are preferred is earthquake prone areas are in Japan.
- Seismic hazard map should give the information about the magnitude of intensity of anticipated earthquakes and seismologist should analyze the pattern of micro seismicity and indicate the possibility of occurrence of earthquakes in potential area.

Earthquake in Chile

The largest earthquake occurred in May 22nd, 1960 in Chile with the magnitude of 9.5 on Richter scale, affecting 90.000 sq. Miles and killing 6,000 people.

Earthquake in Bhuj Town

The earthquake occurred on in Bhuj Town in Gujarat has caused massive damage. Nearly 20.000- 30,000 people were killed and leaving many injured. The intensity of the earthquake is 6.3 on the Richter scale.

Earthquake in Iran

In Iran, an earthquake in 1971 killed 25,000 people, in 1990, 50,000 people were killed and in 2003 many thousands were killed.

4.12.3 Cyclone

Cyclone is a meteorological phenomenon, intense depressions forming over the open oceans and moving towards the land. On reaching the shores, it moves into the interior of the land (or) along the shore lines.

Occurrence: Tropical cyclone in the warm oceans is formed because of heat and moisture. The main requirement of formation of tropical cyclone is that the sea surface temperature must be below 25°C. The tropical cyclones move like a spinning top at the speed of 10-30 km/hour.

In India, cyclones originating from Bay of Bengal are more in number of intensity. But they are relatively less in South West Indian Ocean.

Effect of Cyclone:

- The damage depends on the intensity of cyclone. The damage to human life, crops, roads.
- Cyclone occurrence slows down the developmental activities of the area.

Cyclone Management:

- Satellite images are used by meteorological departments for forecasting the weather conditions, which reveal the strength of intensity of storm.
- Radar system is used to detect the cyclone and is being used for cyclone warming.

4.12.4 Landslides

The movement of earth material like coherent rock, mud, soil and debris from higher region to lower region due to gravitational pull is called landslides.

Causes of Landslides:

- Down hill movement of earth is mainly caused by rain, either increasing the top material weight, lubricating the various layers or making the slope too steep.
- Movement of heavy vehicles on the unstable sloppy regions creates landslides.
- Earthquake, shocks, vibrations and cyclone create landslides.
- Erosion in the hilly tract due to run-off water during rainy period also leads to landslides.

Effects of Landslides:

- Landslides block the roads and divert the passage.
- Erosion of soil increases.

Landslides Management

It is very difficult to control landslides. However, these can be minimized by:

- Unloading the upper parts of the slope.
- Improving the cultivation in the sloppy region, the roots of which provide cohesion and retards the flow of water and its erosion capacity.
- Steepness of the slope can be reduced by developing benches.
- Concrete support can be made at the base of the slope.
- Draining the surface and subsurface water from the weak sloppy regions.

4.12.5 Public Awareness

In terms of the public awareness about impending floods, droughts, earthquakes, cyclones etc., India has achieved widespread but low levels of scattered mixed results. The important activities taken by NDMA include preparation of guidelines and improvements in risk assessment, early warning systems and capacity-building and expanding communication networks.

However, despite of the wide array of activities, there is a widespread lack of public appreciation for disaster preparedness. Moreover, despite a complex structure of disaster management, there is a limited policy interest in disaster risk reduction (DRR). The key areas such as shelter, infrastructure and industry remained largely unfocussed in India.

In our country, there has been a need to establish institutions at all levels to create awareness of disaster and the ways to effectively respond to disasters. Thus, the need for the effective disaster management is building of knowledge base of the local people and strong institutions.

The prevention pays when it is done well. To achieve results, India needs to impart its people the sector-wide awareness about disaster risks. This can be done through the central administration and well-targeted help of the United Nations Development Programme (UNDP), by a series of training activities aimed at the local leaders, institutions and rural women in using the Hyogo Framework for Action (HFA).

While the NDMA has made progress in establishing overall structure for disaster management, several gaps continue to exist in India's disaster preparedness that require focus. One key gap is NDMA's oversight in articulating a list of natural and manmade disasters for which the National Disaster Response Force (NDRF) would apply to. This lack of clarity has left funding options for relief and rehabilitation more ambiguous. Moreover, partially as a result of this gap, so far the guidelines have not been backed up with the resources to implement them. Institutions lack resources and are often incapable in carrying out the necessary activities. Thus, building the capacity of institutions and increasing the flow of resources must become a critical priority of NDMA. It must also aim to raise public awareness through collaboration with non-governmental organizations (NGOs), local government bodies and primarily through panchayati raj institutions.

Different natural calamities can be distinguished from each other in terms of their nature and extent of their impact. Calamities like earthquakes, hailstorms, avalanches, landslides, etc. occur quite suddenly but they are restricted in terms of time and space. Similarly, though floods and cyclones occur with some element of warning yet their occurrence is confined in duration.

Drought, on the other hand, spans over a much longer time-frame and its adverse impact on the economic activities and life of an area is of a more lasting nature. The measures required to meet the threats posed by different calamities, therefore, differ considerably in terms of disaster preparedness and amelioration of the economic and social life of the affected people.

The disaster mitigation works mainly address the following:

- Minimize the potential risks by developing disaster early warning strategies.
- Prepare and implement the developmental plans to provide the resilience to such disasters.
- Mobilize the resources including communication and telemedicine services.
- To help in rehabilitation and post-disaster reduction.

Chapter 5

Human Population and the Environment

5.1 Population Growth

Population Growth and Environmental Issues

The rapid growth of the global population for the past 100 years results from the difference between the rate of birth and death. In 1980, global population was about 1 billion people. It took about 130 years to reach 2 billion. But the population reached to 4 billion within 45 years. Now we have already crossed 6 billion and may have to reach about 10 billion by the year 2050 as per the World Bank calculations.

Causes of Rapid Population Growth

- The rapid population growth is due to decrease in the death rate and increase in birth rate.
- The availability of antibiotics, immunization, increased clean water, food production and air decreases the famine related deaths and infant mortality.
- In agricultural based countries, children are required to help parents in the fields that is why population increases in the developing countries.

Characteristics of Population Growth

- Exponential growth: Now population growth occurs exponentially like 10, 102, 103, 104, etc., which shows the dramatic increase in global population in the past 160 years.
- Doubling time: It is the time required for a population to double its size at a constant annual rate. If a nation has 2% annual growth, its population will double in next 35 years.
- Infant mortality rate: It is the percentage of infants died out of those born in the year. Even though this rate has decreased in the last 50 years, the pattern differs widely in developing countries.

- Total fertility rates (TFR): It is the average number of children delivered by a women in her life time. The TFR value varies from 2 in developed countries to 4.7 developing countries.
- Replacement level: Two parents bearing two children will then be replaced by their off spring. Due to infant mortality their replacement level is changed. But, due to high infant mortality replacement level is generally high in developing countries.
- Male-Female ratio: The ratio of girls and boys should be fairly balanced in a society to flourish. But the ratio has been upset in many countries including China and India. In China the ratio of girls and boys is 100:140.
- Demographic transition: Population growth is generally related to economic development. The death rates and birth rates fall due to improved living condition. This results in low population growth. This phenomenon is referred to as demographic transition.

Problems of Population Growth

- Increasing demands for the food and natural resources.
- Inadequate processing and health services.
- Loss of agricultural lands.
- Unemployment and social political unrest.
- Environmental pollution.

5.1.1 Variation Among Nations

Population Growth Variations

From the dawn of mankind to the turn of the nineteenth century world population grew to a total of one billion people. During the 1800s, human numbers increased at increasingly higher rates, reaching a total of about 1.7 billion people by 1900. World population has grown even more rapidly during the present century, with the greatest gains occurring in the post-World War II period and stands at over 3 times its size in 1900—some 5.9 billion people—today.

Population growth has continued throughout the past three decades in spite of the decline in fertility rates that began in many developing countries in the late 1970s and in some countries, in spite of the toll taken by the HIV/ AIDS pandemic. While the rate of increase is slowing, in absolute terms world population growth continues to be substantial. Global population increase is currently equivalent to adding a new Israel, Egypt, Jordan, West Bank and Gaza to the existing world total each year.

According to Census Bureau projections, world population will increase to a level of nearly 8 billion persons by the end of the next quarter century and will reach 9.3 billion persons, a number more than half again as large as today's total by the year 2050.

The future of human population growth has been determined and is now largely being decided, in the world's less developed nations (LDCs). Ninety-six percent of world population increase now occurs in the developing regions of Africa, Asia and Latin America and this percentage will rise over the course of the next quarter century.

Ninety percent of the world's births and 77% of its deaths will take place in LDCs in 1998. Ninety-nine percent of global natural increase — the difference between numbers of births and deaths—occurs in the developing world. The Census Bureau's projections indicate that early in the next century, crude death rates will exceed crude birth rates for the world's more developed countries (MDC's) and the difference – natural increase—will be negative.

At this point, international migration will become the critical variable determining whether the total population of today's MDC's increases or decreases. These projections show negative natural increase offset by net international immigration through 2019 but, if present trends continue, population of the world's MDC's will slowly begin to decrease from the year 2020 onward.

As the growth rate in the world's more affluent nations becomes negative, all of the net annual gain in global population will, in effect, come from the world's developing countries. This is a listing of the 23 most popular countries in the world. Data are estimates from mid-2006:

- China-1,313,973,713.
- India-1,095,351,995.
- United States-298,444,215.
- Indonesia-245,452,739.
- Brazil-188,078,227.
- Pakistan- 165,803,560.
- Bangladesh- 147,365,352.
- Russia- 142,893,540.
- Nigeria- 131,859,731.
- Japan -127,463,611.

- Mexico-107,449,525.
- Philippines-89,468,677.
- Vietnam-84,402,966.
- Germany-82,422,299.

5.1.2 Population Explosion

The enormous increase in population, due to low death rate (mortality) and high birth rate is termed as population explosion. The human population is not increasing at a uniform rate in all parts of the world.

Causes of population explosion:

- Increase of life expectancy.
- Modern medical facilities reduces death rate & increases birth rate.
- Illiteracy.

Poverty

Infant mortality is one of the most tragic indicators of poverty. There are still 34 developing countries where more than 1 in 10 children die before he or she reaches the age of five. In developing countries few matters opt for smaller families in the face of such high infant mortality rates. Efforts to keep children value and healthy are one of the keys to reduce population growth rates:

- Population explosion leads to environmental degradation.
- Population explosion causes over exploitation of natural resources. Therefore there will be a shortage of resources for the future generation.
- Many of the renewable resources like forest, grass lands, are also under threat.
- The increase in population resources will increase disease economic inequity and communal war.

Remedy: We should immediately reduce the fertility rate through birth control programs.

5.1.3 Family Welfare Programme

Family welfare programme was implemented by the Government of India as a voluntary programme. It is an integral part of overall national policy of growth covering human health, maternity, family welfare, child care and woman's right.

Objectives of Family welfare programming:

- Pressure on the environment due to over exploitation of natural resource.
- Slowing down the population explosion by reducing the fertility.

Population Stabilization Ratio

The ratio is derived by dividing crude birth rate by crude death rate.

Developed Countries

The stabilization ratio of the developed countries is, which is more or less stabilized indicating zero population growth.

Developing Countries

The stabilization ratio of the developing countries nearing 3, which is expected to lower down by 2025. Stabilization in developing countries is possible only through various family welfare programmes.

5.2 Environment and Human Health

Human health and environment are two inseparable entities, if one gets disturbed. Generally a physically fit person, not suffering from any disease, is called a healthy person.

- Nutritional factors.
- Biological factors.
- Chemical factors.
- Psychological factors.

1. Physical Hazardous and their Health Effects:

Physical Hazards	Health Effects
Radioactive radiations	Attacks the cells in the body on the function of gland and organs.
UV radiations	Skin cancer
Global warming	Temperature increase cause famine, mortality.
Chlorofluorocarbons	Damage O3 layer, UV rays cause skin cancer.
Noise	Painful and irreparable damage at human ear.

2. Chemical Hazards and their Health Effects

Chemical Hazards	Health effect
Combustion of fossil fuels: Liberates SO_2, NO_2, O_2 Industrial effluents	Asthma, bronchitis and other lung disease kill cells and cause cancer death.
Pesticides like DDT and Chlorinated pesticides Heavy metals like Hg, Pd, N etc	Affect the food chain contaminate water cause will effect.

3. Biological Hazards and their Health Effect

Biological Hazards	Health effects
Bacteria, viruses and Parasites	Diarrhoea, malaria, parasites, worms, anemia, Cholera etc.

Preventive Measures

- Always wash your hands before sitting for food.
- Cut short and clean your nails.
- Maintaining the skin, teeth, hair.
- Drink chemically treated and filtered water.
- Always boil the food while it is in hot condition.
- Before cooking, wash the raw vegetables and fruits with clean water.
- Try to avoid plastic containers and aluminium vessels.
- Do physical exercise to have proper blood circulation in the body.

5.2.1 Human Rights

Every woman, man, youth and child has the human right to a safe and healthy environment and to other fundamental human rights linked to and dependent upon a healthy environment.

Human Rights relating to the environment are set out in basic human rights treaties and include:

- The human right to a safe and healthy environment.
- The human right to the highest attainable standard of health.
- The human right to ecologically sustainable development.
- The human right to an adequate standard of living, including access to safe food and water.

- The human right of the child to live in an environment appropriate for physical and mental development.
- The human right to full and equal participation for all persons in environmental decision-making and development planning and in shaping decisions and policies affecting ones community, at the local, national and international levels.
- The human right to safe working conditions, including adequate safeguards for pregnant and lactating women.
- The human right to freedom from any type of discrimination.
- The human right to education and information, including information relating to links between health and the environment.
- The human right to share in the benefits of scientific progress.

Work towards sustainable development is increasingly recognizing the importance of a human rights approach. This should not be surprising: the protection of human life in relation to life, health, culture and living standards is central to any social, environmental or economic programme. The right to life cannot be realized without the basic right to clean water, air and land. A human rights approach allows the quality of life of people, in particular the most vulnerable, to be integrated into environmental decision making.

There are two main approaches to human rights and the environment:

- The use of existing human rights,
- The need for new human rights for a safe and clean environment.

Any discussion of environmental human rights must encompass three areas of work:

- The right to a clean and safe environment.
- The right to act to protect the environment.
- The right to information and participate in decision-making related to environment.

Human Rights Features

Universal declaration of human rights (UNDHR) by the UN was established in 1948. Some of the main declarations of human rights, which are globally accepted are as follows:

1. Human Right to Freedom

- Every citizen has the freedom to express his views freely.
- Citizens can assemble at any place to express their views.

- They have freedom to form unions or association.
- They have freedom to build their houses wherever they like.
- They have felt right to start any profession.

2. Human Right to Property

- Every human being has the right to earn property.

3. Human Right to Freedom of Religion

- Every citizen has the freedom to choose their religion according to his wishes. All religions are equal before the law.

4. Human Right to Culture and Education

- All the citizen has equal both in culture and education. The minority communities like Christian, Muslims have their own rights to conserve the culture, language and to establish educational institutions of their own choice.

5. Human Right to Constitutional Remedies

- A citizen is denied in any of three fundamental rights. He or She can go to the court for protection. The court has the power to protect the basic rights to the citizens.

6. Human Right to Equality

- All citizens are equal before the law. There is no any discrimination on grounds of religion case, sex or place of birth. All are given equal opportunity for employment.

7. Human Right Against Exploitation

- Children should not be employed as labors. Every citizen has the right to fight against exploitation.

8. Human Right to Food and Environment

- All human beings have the right to get sufficient healthy food, safe drinking water and healthy environment.

9. Human Right to Good Health

- All the human beings have the right to have very good physical and metal health.

5.2.2 Value Education

Value Education is an instrument used to analyze our behavior and provide proper direction to our youths. It teaches them the distinction between right and wrong, to be compassionate, helpful, loving, generous and tolerant. So that a youth can move towards the sustainable future.

Example:

If a person is highly qualified, well employed and rich living style, sometimes he does not know how to behave properly with his environment. But, value education does teach him this.

Objectives of Value:

- To improve the integral growth of human being.
- To create attitudes and improvement towards sustainable lifestyles.
- To increase awareness about our national history, our cultural heritage, constitutional rights, national integration, community development and environment.
- To create and develop awareness about the values.
- To understand about our natural environment in which how land, air and water are interlinked.

Types of Values

- Universal Values (or) Social Values: Universal values tell us about the importance of the human conditions. These values are reflected in life, joy, love, truth, etc.
- Cultural Values: Cultural values varies with respect to time and place. These values are connected with right and wrong, good and bad, true and false and behavior of human beings. It is reflected in language, education, law, etc.
- Individual Values: Individual values are our personal principles and the result of individual personality and experiences, parents and teachers are the main role to shape our individual value. It is reflected in individual goals, relationships, etc.
- Global Values: Global values stress the concept that the human civilization is a part of the planet and similarly nature and natural phenomena over the earth are interconnected are interlinked with special bonds of harmony. If this harmony is disturbed anywhere there will be an ecological imbalance leading to catastrophic results.

- Spiritual Values: Spiritual values promote conservation and transform our consumer approach. It is reflected in self-restraining, self-discipline, contentment, reduction of wants, etc.

5.3 HIV/AIDS

HIV

AIDS is the abbreviated form for Acquired Immune Deficiency Syndrome caused by a virus called HIV (Human Immune deficiency Virus). Many myths have been spread about AIDS and it is very important for everyone to know the facts about HIV and AIDS.

Origin of Hiv/Aids

AIDS was discovered in 1983. Though sufficient knowledge has been gained about the disease, yet a definite source of this virus could not be identified. But the following theories have been suggested.

Through African Monkey:

Most of the evidence have suggested that the AIDS has spread from Africa. It has been believed that one HIV has transferred to humans from African Monkey or Chimpanzees.

Through Vaccine Programme:

- HIV has spread in Africa through HIV contaminated polio vaccine, prepared from monkey's kidney.
- It had spread through hepatitis-B viral vaccine in LOS Angles and New York.
- It had also spread through small pox vaccine programme of Africa.

Factors Influencing Modes of Transmission of HIV

HIV is spreading due to the following activities:

- HIV from infected person can pass to a normal person through blood contact, generally during unprotected sex with infected person.
- Using needles or syringes, contaminated with small quantities of blood from HIV positive person, also transmit HIV to others.
- HIV can also pass from infected mothers to their babies during pregnancy, delivery or breast feeding.
- Blood transfusion from the infected person, at the time of accidents or pregnancy also results in HIV/AIDS.

- Women are more vulnerable to HIV Biologically the male to female transmission of HIV is 2 to time more efficient than female to male transmission.
- Women around 18-20 years are at more risk, since their critical tissue is more vulnerable to invading HIV.
- Since the majority of HIV infections occur in women of child-bearing age, transmission of HIG of their new born babies happen easily.

Factors not Influencing Transmission of HIV

HIV is not spreading by the following activities:

- Tears, food and air, cough, handshake, mosquito, flies, insect, bites, urine, saliva during normal kissing, sharing of utensils, clothes, toilet, bathroom, etc.
- Functions of HIG in Human Body: AIDS itself does not kill humans. The person dies as the disease weakens the human immune system, mainly responsible to fight diseases.
- White Blood cell (NBC), responsible in the formation of antibodies, are called T-helper cells. T-helper cells are the key infection fighters in the immune system. The HIV enter into the human body and destroys the T-cells, as a result of which various types of infection diseases occur. Even cancer can easily develop in the HIV infected persons.

Symptoms (or) Diagnosis of HIV/AIDS

Many people have no symptoms. When they are first infected with HIV. But some people get fever, headache, and fatigue. During this time, HIV is present in large amounts in semen and vaginal fluids and it is very easy to pass the infection to another person.

HIV is very active inside a person's body. The virus multiples and kills more and more T-cells of the immune system.

Consumption of alcohol increases the susceptibility to infection and progress of AIDS.

Minor Symptoms

- Persistent cough for more them one month.
- General skin diseases.
- Viral infection.
- Fungus infection in mouth and throat.
- Frequent fever, headache, and fatigue.

Major Symptoms

- Fever for more than one month.
- Diarrhea for more than one month.
- Cough and TB for more than 6 months.
- Fall of hairs from the head.
- 10% of the body weight gets reduced within a short period.

Control and Preventive Measures of AIDS: Once a person is infected with HIV the person remain infected for life. There is neither a cure nor a vaccine but can be prevented. "Prevention is better than cure" is a slogan for all of us.

The basic approaches to control AIDS are:

- Education: Health education enables people to avoid indiscriminate sex and encourages the use of condoms. One should avoid the use of sharing razors, needles and syringes.
- Prevention of Blood Borne HIV Transmission: People in high risk groups should not donate blood. Blood should be screened for HIV before transmission and strict sterilization practices should be followed in hospitals.
- Primary Health Care: AIDS awareness programme should be encouraged. Voluntary health agencies should participate in large. Training Programme to doctors and paramedical workers should be conducted.
- Counseling Services: Counseling services should be provided either in person or through telephone.
- Deng Treatment: Testing HIV positive does not mean for the end. They can still stay healthy leading productive lives for many years. Seeking early medical care and staying active are very vital in managing HIV. The immune system has to be kept boosted by taking nutritious diet and maintaining a stress-free mind.

Effects of AIDS/HIV

- Large number of death occurs, which affect environment and natural resources.
- Due to large number of deaths, there is loss of labor and level of production decreases.
- More water is required for maintaining hygiene in AIDS affected locality.
- The people affected by HIV, cannot perform work well due to lack of energy and frequency fever and sweating.

5.4 Women and Child Welfare

Women and Child are usually soft, who suffer in a number of ways mainly because they are weak, helpless and economically dependent.

Women Welfare

The main aim of women welfare is to improve the status of the women by providing opportunities in education, employment and economic independence.

Need of Women Welfare

Generally women face the following problems in the society, so there is an urgent need for policy reforms and more stringent legislation, education and legal awareness among women for checking injustice towards her.

- Generally women suffer gender discrimination and devaluation at some, at work place, in matrimony, in public life and power.
- High number of cases of dowry deaths, rape, domestic violence, criminal offenses and mental torture to women.
- The human rights of women are violated, in the male dominated society.
- Generally in policy making and decision making process, women are neglected.

Objectives of Women Welfare

To overcome the above problems, a sound national strategy is needed with the following objectives:

- To provide education.
- To impart vocational training.
- To generate awareness about the environment.
- To improve the employment opportunities.
- To aware problems of population.

Objectives of a National Commission for Women

A National Commission for women has been created by Government of India, its main objectives are:

- To examine constitutional of legal rights for women.

- To review existing legislation's.
- To sensitive the enforcement and administrative machinery to women causes.

Environmental Degradation and Women Welfare

The developmental work not only affects the natural environment but also affects the traditional, social, cultural and family life of women.

Example:

After losing the forest cover and getting rehabilitated from their native places, men usually migrate to towns for some job while the women are left behind the look after their family with little resources.

Various Organizations towards Women Welfare

- The National Network for Women and Mining (NMWM). It is fighting for a "gender audit" of India's mining companies.
- United Nations Decade for Women. It witnessed inclusion of several women welfare related issues on international agenda.
- International convention on the Elimination of all Forms of Discrimination Against Women (CE DAW). It has created an international standard for the protection and promotion of women human and social-economic up lift-ment.
- Non-Government Organizations (NGO's) As Mahila Mandals. It creates awareness among women of remote villages to empower them, train them, educate them and help them to become economically self-dependent.

Child Welfare

Children occupy nearly 40% of total population. They are considered to be the assets of a society of 21 million children born every year in India, 20 million children in our country are estimated to be working as child labors in various hazardous industries like match industry, fire work industry.

Reason for Child Labors

- Poverty: Poverty is the main reason to force these children to work in unhealthy conditions.
- Want to Money: Parents require money for their family, so they are in a position to send their children for work.

Various Organizations Towards Child Welfare

Un-Conventions on Rights of Child or International Law

It formulated a set of international standards and measures to promote and protect the well being of children in our society.

Rights of the Child

The international law defines right of the child to survival, participation, development and protection:

- The right to survival: It emphasizes on good standards of living, good nutrition and health.
- The right to participation: It means freedom of thought and appropriate information to the child.
- The right to development: It ensures access to education, childhood and care and support, social security and recreation.
- The right to protection: It means freedom from exploitation, inhuman treatment and neglect.

World Summit on Children

It had focused agenda for the wellbeing of the children targeted to be achieved in the beginning of the new millennium.

Ministry of Human Resource Development (MHRD)

It concentrates on child's health, education, nutrition, clean and safe drinking water, sanitation and environment.

Environment Degradation and Child Welfare

Children are most affected due to environmental pollution. Water borne diseases are the biggest threat to children. Around 6 million children are affected by this disease in India. Even the child growing in the mother's womb, gets affected by environmental toxins.

Centre of Science and Environment (CSE)

Its scientific report says, "Children consumer more water, food and air than adults and hence more susceptible to any environmental contamination." So, it is essential to keep the cleaner environment to our children for the better and healthy life.

5.5 Environmental Impact Analysis

Environment Impact Assessment or EIA can be defined as the study to predict the effect of a proposed activity or project on the environment. A decision making tool, a EIA compares various alternatives for a project and seeks to identify the one which represents the best combination of the economic and environmental costs and benefits.

EIA systematically examines both the beneficial and adverse consequences of the project and ensures that these effects are taken into account during the project design. It helps to identify the possible environmental effects of the proposed project, proposes measures to mitigate the adverse effects and predicts whether there will be significant adverse environmental effects, even after the mitigation is implemented.

By considering the environmental effects of the project and their mitigation early in the project planning cycle, the environmental assessment has many benefits, such as protection of the environment, optimum utilization of resources and saving of time and cost of the project. Properly conducted the EIA also lessens the conflicts by promoting the community participation, informing decision makers and helping lay the base for environmentally sound projects. Benefits of integrating EIA have been observed in all the stages of a project, from exploration and planning, through construction, decommissioning, operations and beyond site closure.

Evolution of EIA

EIA is one of the successful policy innovations of the 20th Century for environmental conservation. Thirty-seven years ago, there was no EIA but today, it is a formal process in many countries and is currently practiced in more than 100 countries. EIA as a mandatory regulatory procedure originated in the early 1970s, with the implementation of the National Environment Policy Act (NEPA) 1969 in the US.

A large part of the initial development took place in a few high-income countries, like Australia, Canada and New Zealand. However, there were some of the developing countries as well, which introduced EIA relatively early - Columbia (1974), Philippines (1978).

The EIA process really took off after the mid-1980s. In 1989, the World Bank adopted the EIA for major development projects, in which a borrower country had to undertake an EIA under the Bank's supervision.

Table: Evolution and History of Eia

Periods	Development of EIA
Pre-1970	Projects review is based on the technical/engineering and economic analysis. Limited consideration given to the environmental consequences.

Early/mid – 1970s	EIA was introduced by NEPA in 1970 in US. Basic principle: Guidelines, procedures including the public participation requirement instituted. Standard methodologies for impact analysis developed (e.g. Checklist, matrix and network). Canada, New Zealand and Australia became the first countries to follow NEPA in 1973-1974. Unlike the Australia, which legislated EIA, Canada and New Zealand established administrative procedures. Major public inquires help shape the process's development.
Late 1970 and early 1980s	More formalized guidance. Other industrial and developing countries introduced the formal EIA requirements (France, 1976; Philippines, 1977), began to use the process informally or experimentally (Netherlands, 1978) or adopted elements, such as the impact statements or reports, as part of development applications for planning permission. Use of EA by the developing countries (Brazil, China, Philippines, Indonesia) Strategic Environment Assessment (SEA), risk analysis included in EA processes. Greater emphasis on the ecological modeling, prediction and evaluation methods. Provision for public involvement. Coordination of EA with the land use planning processes.
Mid 1980s to end of decade	In Europe, EC Directive on EIA establishes the basic principle and procedural requirements for all member states. Increasing the efforts to address cumulative effects. World Bank and other leading international aid agencies establish the EA requirements. Spread of EIA process in Asia.
1990s	Requirement to consider the trans-boundary effects under Espoo convention. Increased use of GIS and other information technologies. Sustainability principal and global issues receive increased attention. India also adopted the EIA formally. Formulation of EA legislation by many developing countries. Rapid growth in EA training.

History of EIA in India

The Indian experience with the Environmental Impact Assessment began over 20 years back. It started in 1976 to 1977 when the Planning Commission asked the

Department of Science and Technology to examine the river-valley projects from an environmental angle. This was subsequently extended to cover those projects, which required the approval of the Public Investment Board. Till 1994, the environmental clearance from the Central Government was an administrative decision and lacked legislative support.

On 27 January 1994, the Union Ministry of Environment and Forests (MEF), Government of India, under the Environmental Protection Act 1986, promulgated an EIA notification making Environmental Clearance (EC) mandatory for expansion or modernization of any activity or for setting up new projects listed in Schedule 1 of the notification. Since then there have been 12 amendments made in the EIA notification of 1994.

The MEF recently notified new EIA legislation in September 2006. The notification makes it mandatory for various projects such as mining, river valley, thermal power plants, infrastructure (road, ports, highway, harbours and airports) and industries including very small electroplating or foundry units to get the environment clearance. However, unlike the EIA Notification of 1994, the new legislation has put the onus of clearing projects on the state government depending on the size/capacity of the project.

Certain activities permissible under the Coastal Regulation Zone Act, 1991 also require similar clearance. Additionally, the donor agencies operating in India like the World Bank and the ADB have a different set of requirements for giving the environmental clearance to projects that are funded by them.

The EIA Process

The stages of an EIA process will depend upon the requirements of the country or donor. However, the most EIA processes have a common structure and the application of the main stages is a basic standard of good practice.

The environment impact assessment consists of eight steps with each step equally important in determining the overall performance of the project. Typically, the EIA process begins with screening to ensure the time and resources are directed at the proposals that matter environmentally and ends with some form of follow up on the implementation of the decisions and actions taken as the result of an EIA report. The eight steps of the EIA process are presented in brief below:

- Screening: First stage of EIA, which determines whether the proposed project, requires an EIA and if it does, then the level of assessment is required.
- Scoping: This stage identifies the key issues and impacts that should be further investigated. This stage also defines the boundary and time limit of the study.

- Impact analysis: This stage of EIA identifies and predicts the likely environmental and social impact of the proposed project and evaluates the significance.
- Mitigation: This step in EIA recommends the actions to reduce and avoid the potential adverse environmental consequences of the development activities.
- Reporting: Reporting presents the result of EIA in a form of a report to the decision-making body and other interested parties.
- Review of EIA: It examines that the adequacy and effectiveness of the EIA report and provides the information necessary for decision-making.
- Decision-making: It decides whether the project is rejected, approved or needs more changes.
- Post monitoring: This stage comes into play once the project is commissioned. It checks to ensure that the imVVpacts of the project do not exceed the legal standards and implementation of the mitigation measures are in the manner as described in the EIA report.

The overview of the EIA process is represented in figure:

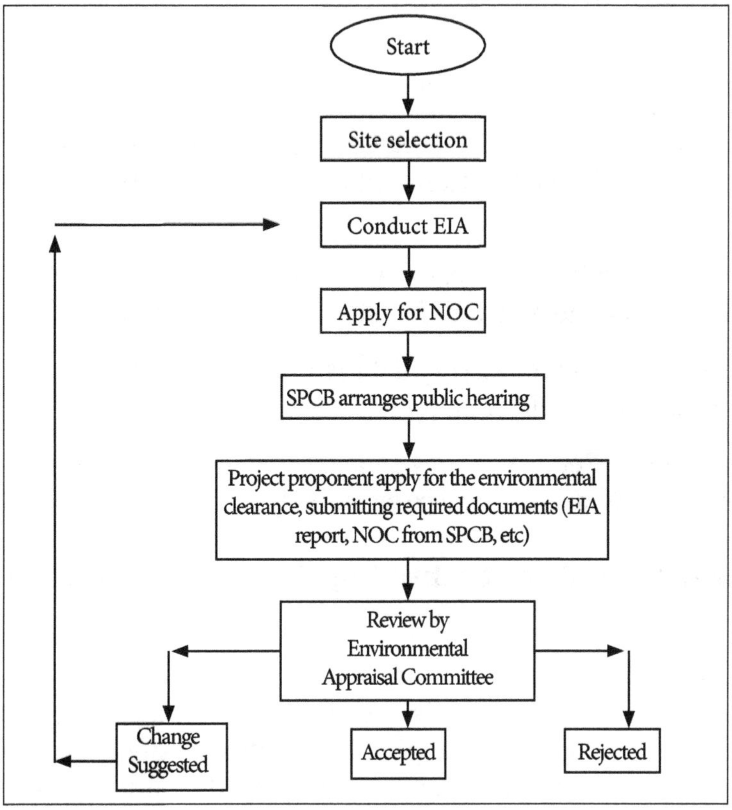

Generalized process flow sheet of the EIA process.

5.6 GIS: Remote Sensing

Geographical Information System

GIS is a technique of superimposing various thematic maps using digital data on a large number of inter related aspects.

Application of GIS

- Interpretations of polluted zones, degraded lands can be made based on GIS.
- GIS can be used to check unplanned growth and be made based on GIS.

Remote Sensing

Remote sensing refers to any method, which can be used to gather information about an object without actually coming in contact with it. At present the term remote sensing is used more commonly to denote identification of earth feature by detecting the characteristics electromagnetic radiation that is reflected /emitted by the earthy.

Example:

The remote sensing image of land can be used to derive information of vegetative cover, water bodies, land use, etc.

Application of Remote Sensing

- In Agriculture: In India, the agriculture sector sustains the livelihood of around 70% of the population and contributes to about 35% of the net national product.
- In forestry: Sustainable forest management requires reliable information on the type, density and extent of forest cover, wood volume and biomass.
- In land cover: Remote sensing data is covered to map, the spatial resolution plays a role on the scale of mapping.
- Water resources: Remote sensing data has been used in many applications related to water resources such as surface water body mapping, wet land and many more.

5.6.1 Role of Information Technology in Environment and Human Health

Information technology has tremendous potential in the field of environment education and health as in any other field like economics, business, politics or culture. Development of internet facilities, Geographic Information System (GIS) and information through the satellites has generated a wealth of up-to-date information on various aspects of environment and health.

A number of software have been developed for the environment and health studies which are used friendly and can help an early learner in understanding the subject.

Database on Environment System

Database is the collection of the interrelated data on various subjects. It is usually in computerized form and can be retrieved whenever required. In the computer the information of database and can be very quickly retrieved. The comprehensive database includes the forest cover database, wildlife database, conservation database, etc. database is also available for diseases like Malaria, HIV/AIDS, Fluorosis, etc.

National Management Information System (NMIS)

NMIS of the Department of Science and Technology has compiled a database on Research and Development Projects along with information about research scientists and personnel involved.

Environmental Information System (ENVIS)

- The Ministry of Environment and Forests, Government of India has created the information System called Environmental Information System (ENVIS). With its headquarters in Delhi, it functions in 25 different centres all over the country.
- The ENVIS centres work for generating a network of database in areas like pollution control, clean technologies, coastal ecology, remote sensing, biodiversity, media related to environment, western Ghats and eastern environmental management, renewable energy, desertification, mangroves, Himalayan ecology, wildlife, mining etc.

Remote Sensing and Geographical Information System (GIS)

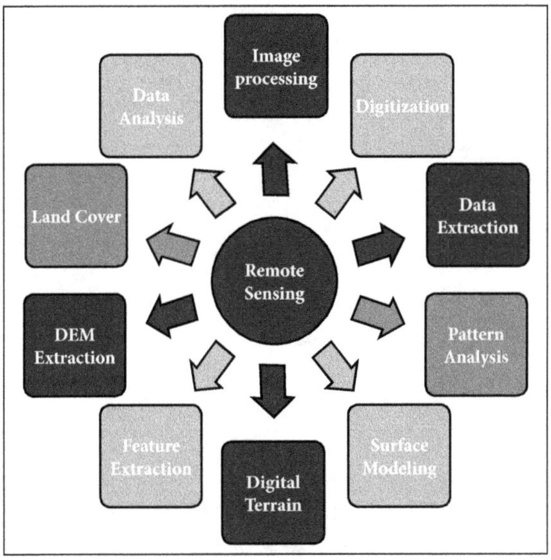

Function of remote sensing.

Satellite imageries provide us the actual information about various physical and biological resources and also to some extent about their state of degradation in a digital form through remote sensing. We are able to gather the digital information on environment aspects like water logging, deforestation, desertification, urban sprawl, river and canal network, mineral and energy reserves and so on.

Geographical Information System (GIS)

GIS has proved to be a very effective tool in the environmental management. GIS is a technique of superimposing various thematic maps using the digital data on a large number of inter-related or inter dependent aspects. Several useful software have been developed for working in the field of GIS.

Different thematic maps containing the digital information on a number of aspects like water resources, human settlements, industrial growth, road network, forest land, soil type, crop land or grassland etc., are superimposed in a layered form in computer using software.

Such information of polluted zones, degraded lands or diseased cropland etc. can be made based on the GIS. Planning for locating the suitable areas for industrial growth is now being done using the GIS by preparing Zoning Atlas. GIS serves to check unplanned growth and helps in providing correct, success of conservation efforts, reliable and verifiable information about forest cover etc.

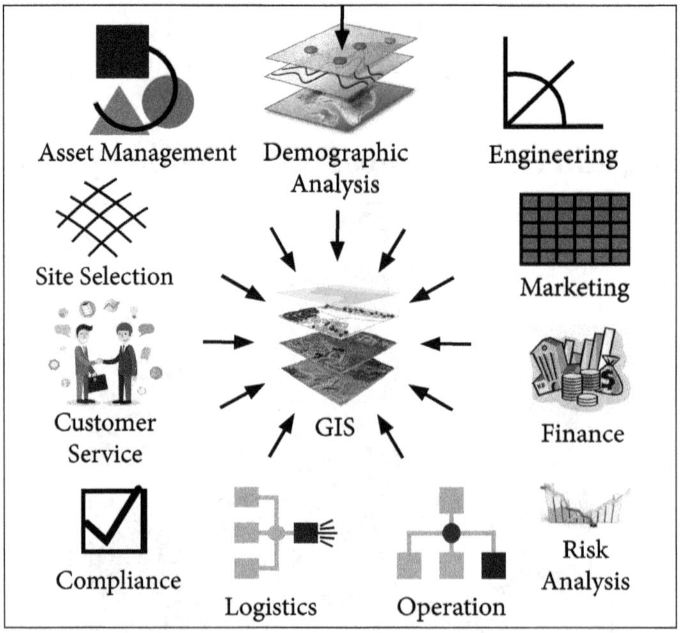

GIS Operation.

They also provide the information of atmospheric phenomena, like approach of monsoon, ozone layer depletion many new reserves of oil, minerals, etc. with the remote sensing and GIS play a key role in resource mapping, management, environmental conservation, planning and environmental impact assessment.

It also helps in identifying the several disease infested areas which are prone to some vector-borne diseases like malaria, schistosomiasis etc. based upon mapping of such areas. There are several Distribution Information Centres (DICs) in our country that are linked with each other and with the central information network having access to the international database.

They also provide the information of atmospheric phenomena like approach of monsoon, ozone layer depletion, smog, inversion phenomena, etc. We are able to discover many new reserves of oils, minerals etc. with the help of information generated by remote sensing satellites. Thus the remote sensing and GIS play a key role in environmental conservation, resource mapping, management and planning and environmental impact assessment.

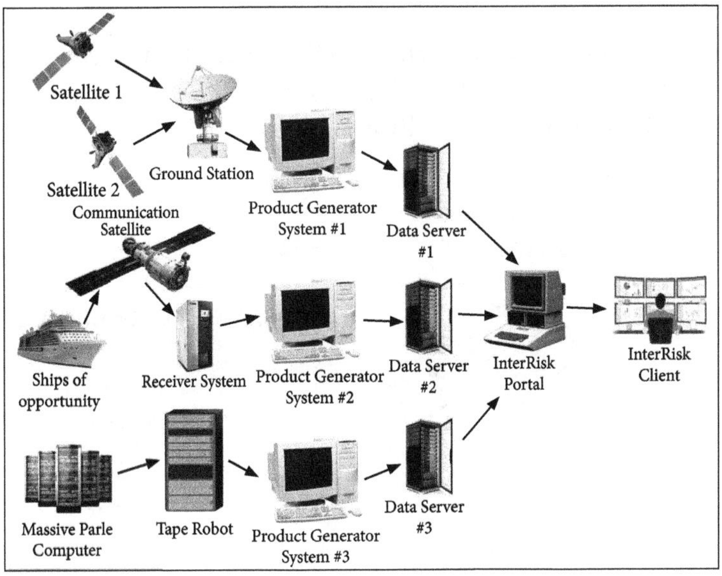

Satellite data processing.

The World Wide Web

With resources material on every aspect, class-room activities and digital files of photos, animations, web-exercises, power-point lecture presentations and quiz has proved to be extremely useful for both the students and the teachers of environmental studies.

5.6.2 Case Studies

The exponential growth in urban population has impacted energy consumption not only in Bangalore but in the rest of India as well. India and Bangalore specifically will need to concentrate on various sectors of energy production in order to meet their rapidly growing demand.

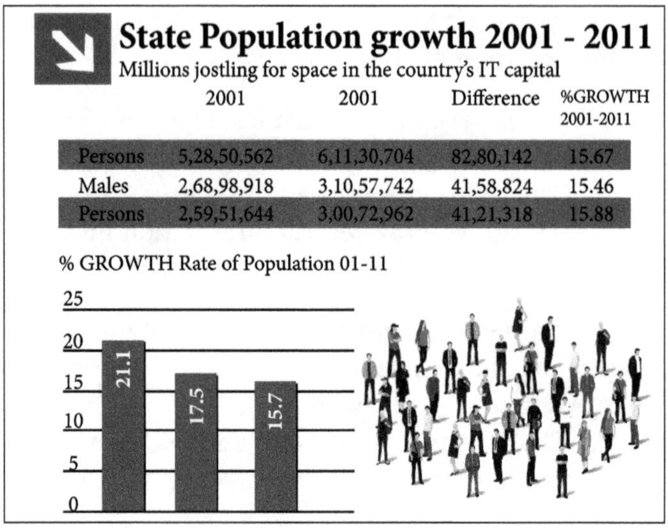

Karnataka Population Growth, 2001-2011.

Although Karnataka, the southwestern state that includes the city of Bangalore, has a population density that is less than the national average, Karnataka's population density has grown by more than 8 million people between 2001 and 2011.

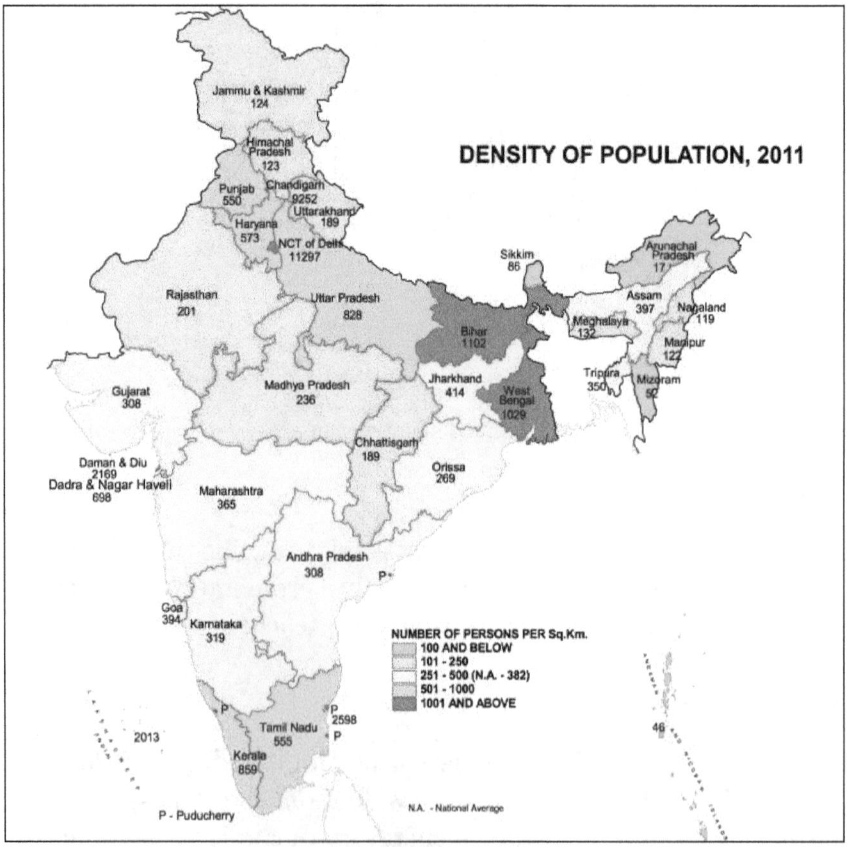

Population Density in 2011.

For American's it was Thomas Friedman who made Bangalore famous. This city of just over eight million residents is often called the Silicon Valley of India, its technology parks and outsourcing services the driving force behind the country's remarkable recent economic boom. For Friedman, Bangalore was the key to understanding the new global economy and he came up with a snappy catchphrase to describe it, which in turn became the title of the best-selling book "The World is Flat". In this flat new world, India's "knowledge economy" would rescue millions of Indians from rural poverty and usher them into a world of eight percent growth rates and abundant clean energy.

The authors of this case study traveled to Bangalore in March 2012 and what we found was similar, yet different from what Thomas Friedman had in mind. In many ways, it was more exciting but life also seems much more challenging for the average Indian. There is no doubting India's sincerity about shifting, over time, to a low-carbon future. All has to do is look at the number of bicycle rentals and associated clean energy billboards one finds when walking around Bangalore.

However, the vision of that future that we found in Bangalore will demand a radical change in the mindset of the government and of those who have done most to create the Friedman's Flat World. In this way, creating a new clean energy economy, whose success or failure will be critical in determining the fate of India as well as the planet.

Bangalore's Clean Energy Billboards, March 2012.

Introduction to Bangalore

Location

Bangalore is situated in southwestern India in the state of Karnataka on the Deccan Plateau at approximately 3000 feet above sea level. Bangalore's relatively high altitude allows the city to experience a milder climate than many large cities situated on the coasts or at lower elevations. This city is located at the ridge top of the main water parting the Arkavathi and the South Pinakini Rivers.

Recent Population Growth

Rapid industrialization and globalization have led to a surprising rise in population and expansion of Bangalore's city limits. Recently, Bangalore became the third largest city in India. According to the 2001 census, the population of Bangalore was 56.86lakhs. Bangalore grew at a much faster rate and the population of Bangalore increased from 41.30lakhs to 56.86lakhs during the decade 1991 to 2001. Most recently, the 2011 Indian census figures were released in the spring of 2012. Bangalore now has a population of 9.59 lakhs and is the fastest growing large urban area in India. The 46.68% growth rate in the last decade, 2001-2011, continues to surprise government officials and planners.

Bangalore's Economic Importance

Bangalore has a strong and balanced economy represented by light and heavy engineering, automobiles, earthmoving, aeronautics, textiles and high technology. The United Nations Human Development Report 2001 ranked the city fourth along with Austin (USA), San Francisco (USA) and Taipei (Taiwan) as the top "Technology Hubs of the World."

Public Sector Undertakings and the textile industry initially drove the Bangalore's economy, but the focus in the last decade has shifted to high-technology service industries. Bangalore's US$ 47.2 Billion economy makes it a major economic center in India and as of 2001 Bangalore's share of US$ 3.7 Billion in foreign direct investment (FDI) made it the 3rd highest recipient of FDI for an Indian City.

With over 103 Central and State research and development institutions, Indian Institute of Science, National Law School of India, 45 Engineering Colleges, world class health care facilities, medical colleges and institutions and a host of other institutional infrastructure, Bangalore is a much sought-after destination for education and research. Bangalore has also enjoyed a favorable positioning that has created job opportunities and rising income levels in excess of population growth.

Bangalore's information technology (IT) industry is divided into 3 main "clusters" such as Software Technology Parks of India, Bangalore (STPI), International Technology Park Ltd. (ITPL) and Electronics City. Infosys and Wipro, India's 2nd and 3rd largest software companies are headquartered in Electronics City. Today, Bangalore is home to 66 Fortune 500 companies, 682 multi-national corporations (MNCs), 1,685 IT/ITES and 131 Biotech companies. Bangalore accounts for half of the approximately 260 biotechnology companies in India. Biocon, headquartered in Bangalore, is the nation's leading biotechnology company and ranks 16th in the world in revenues. Jack F. Welch Technology Research Center in Whitefield, Bangalore is the 2nd largest research facility of General Electric (GE) in the world. The Research Center employs around 1600 people, including more than 1000 doctorates.

Bangalore's Population and Natural Resources Dilemma

Boom in the IT sector, the ever increasing migratory population and the added natural increase in the population has increased the pressure on the natural environment and infrastructure. Spurred by economic development and the space technologies revolution, rapid urban growth and restructuring are expected over the next 10 to 20 years (Masser and Ottens, 1999). This will be characterized by urban redevelopment in the city center, suburban expansion with new sub-centers and leap-frog urban sprawl.

Urban growth has two contradictory facets. On the one hand, mega-cities act as engines of economic and social growth. On the other, most of this is accompanied by both the poverty and environmental degradation, e.g. encroachment on the valuable agricultural land, increasing use of the private car and energy consumption, inner city decline, premature write-down and under-utilization of the existing built environment. Impacts of land use changes on environmental sustainability will become globally significant through their cumulative effects. This is considered to be one of the major global change issues (Vitousek, 1994).

Faced with severe negative impacts, urban planners need to rethink the most important development policies and manage the urban sprawl and urban growth on a more sustainable basis. For instance, in the USA, urban sprawl has sparked off a national debate over land use policy that includes smart growth management and growth boundary measures (Bruecker, 2000).

Growth brings prosperity to the citizens and improves the standard of living. At the same time, growth places greater strain on basic infrastructure and services, which have not been designed to cope with such rapid growth. In many instances, economic growth may also not be equitable and may create the islands of prosperity and poverty. If rapid urban growth is not sustainable, then what impacts are placed on the natural resources and the environment. In developing countries such as India (Than gavel, 2000) and China (Yeh and Li, 2001), patterns of urban growth have been studied in the context of their special social and economic circumstances.

Pressure on Natural Resources

1. Water Supply and Storm Water Drainage

Presently, the water supply system for Bangalore is adequate at current usage levels. However, the draw distances are problematic given the city's population projections. Water distribution is uneven with areas inside the city limits receiving adequate supplies and peripheral areas experiencing the water stress. Due to the uneven growth and leap-frog development, water amenities could not be extended. This causes serious concerns for the planners.

Drainage during rain events is another area of concern. Due to rapid population growth, natural drainage systems are being built on and even lakes in the metropolitan area are filled in and develop. Topography cannot be altered, so during the unexpected heavy rains, Bangalore experiences urban floods on a regular basis. This situation causes increased stream pollution, which can lead to serious sanitation and health issues.

2. Solid Waste Management

Development around the world provides for some unique challenges. In Bangalore, one of these challenges is solid waste management. As the population has increased, the stress on the infrastructure of the city has led to an increase in all types of waste.

Collection and transportation coverage is very good per government reports. However, there is a wide gap between official reports and reality. Proper and adequate treatment or disposal facilities are actually lacking. Only 10% of waste is recycled because there is only one recycling facility in the city. There are six city authorized landfills, which is not enough to handle the 30,000 tons of waste generated each day.

During rain events, these landfills become plunge pools for breeding mosquitoes and radiate diseases to the surrounding areas. Litter and large piles of trash on the city streets are also an environmental and health hazard during heavy rains due to the inadequate storm drainage system of Bangalore. The resulting discrepancy between solid waste management systems and the growing need for expanded collection and disposal facilities has left an accumulating amount of solid waste with the urban environment producing unsanitary conditions.

Solid Waste in the Urban Environment.

3. General Waste Management

As the city of Bangalore continues to grow, waste management moves to the forefront of concern for many. Construction of new businesses and housing provides jobs for many and waste that needs to be disposed of. The city government of Bangalore (BBMP) recognizes waste management is a major concern. There have been a number of laws passed in the recent years, however, they are difficult to enforce. For example, construction sites can be cited if caught leaving debris on sidewalks. Fines can only be levied if the person is caught. So there is no deterrent nor has benefit for construction sites to dispose of waste in any of the new 20 unused quarries the city set up for specifically construction disposal purposes.

A Local Park in Bangalore.

Zero Waste Policies are very popular as Bangalore continues to grow. All waste is divided for either segregation or composting. The city government does sponsor some programs to encourage people to participate in door-to-door daily collection, but these programs are often mismanaged and unorganized. There is often inconsistent pickup or residents don't see the true value in segregating waste.

In an article from The Deccan Herald, (October 23, 2011) the Karnataka Panchayat Raj Act of 1994 was evaluated and criticized. Although the act, written 18 years ago, set forth some standards for general waste management, the cost of these programs was left to the Gram Panchayats that did not have the money to follow through. So, through re-evaluation of this Act, some new goals have been determined. Toilets will be built in certain locations to help eliminate public defecation. All building owners who rent to businesses or residents must provide toilet facilities for those using the building. All public restrooms must have a regular water supply in order to flush the system. For new construction, door numbers will be issued only if there is a toilet. These are only a few of the guidelines the Dakshina Kannada Zila Panchayat has issued. Workers working to comply with these goals will be provided the proper gear and equipment.

4. Air Pollution

Rising traffic congestion is a serious issue confronting Bangalore. The combination of population and economic growth of Bangalore contributes to the increased number registered vehicles and miles/kilometers driven. Added to this problem is the fact that there is no rail-based commuter system, which only compounds the situation.

5. Land Use Cover/Change: Loss of Bio-Diversity and Prime Agricultural Lands

The city has a tradition of being a "Garden City" with plenty of green spaces and water bodies. However, the very high growth rate in the past two decades is having an adverse impact on the urban canopy, open spaces and lakes. The greenery loss takes place at two levels. One is the loss of tree groves and forests. This mainly happens when there infrastructure development activities such as extension of roads, etc. Next is the conversion of land due to speculation and monetary gains by more intensive use of land.

These two main factors account for dramatic loss of vegetation and agricultural lands on the urban fringe. Population growth and the lack of land inside the city for development purposes have significantly increased land values close to the urban core. Developers then turn to less expensive land on the fringe to build offices. The city government is encouraging developers to build vertically, however economic forces leading to leapfrog development and sprawl are too strong.

6. Energy Consumption

Energy has been universally recognized as one of the most important inputs for economic growth and human development. There is a strong two-way relationship between economic development and energy consumption. On one hand, growth of an economy, with its global competitiveness, hinges on the availability of cost-effective and environmentally benign energy sources and on the other hand, the level of economic development has been observed to be dependent on the energy demand (EIA, 2006). It seems clear that there is a positive relationship between total primary energy consumption to GDP, population growth and per capita energy consumption. However, a negative relationship does exist between the energy use and the production of the energy resources in the case of India. One can summarize that total primary energy use is one of the key components of the GDP. Population growth is an important factor for the total primary energy consumption and one of the major contributors for the demand of more energy resources. Per capita energy consumption in the economy has a positive relationship with the total energy use and hence it is one of the important factors of the total energy consumption. As evidence from the developing countries, such as India, the more the country develops economically, the demand for energy resources also increases. An analysis of data also suggests in the same manner that an increase in GDP, Population and Per capita consumption leads to more demand of energy resources.

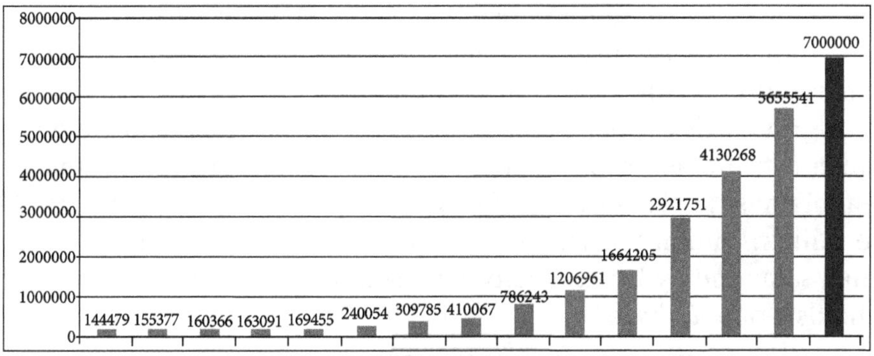

Population Growth of Bangalore City, 1871–2007*.

*The population for 2007 is an estimate.

Table: Comparison of population, economic indicators and resource consumption.

	CHINA	INDIA	USA	EURO
Population (millions)	1288	1064	291	306
GNP ($- billions)	1417	571	11,013	6,978
Water Use (cubic kilometers)	526	500	467	185
Energy Use (quadrillion BTUs)	52	23	97	50
GNP/Pop (per capital income)	1100	536	37,844	22,803
Water/GNP (cubic meters water per $ 1.00 of GNP)	0.37	0.88	0.04	0.03
Energy / GNP (BTUs per $1.00 of GNP)	36717	39,926	8,808	7,210
Water/PoP (par capital water use in cubic meters)	408	470	1,606	605
Energy / PoP (per capital energy use in millions of BTUs)	40	21	333	164

Exponential growth in population represented by the graph above has impacted energy consumption not only in Bangalore but also in the rest of India as well. Likewise, as seen above, India's energy usage per GNP is clearly the highest in the world when compared to the U.S., China and the countries of the EU. India boasts a growing economy and is increasingly a significant consumer of oil and natural gas.

According to the International Energy Agency (IEA), hydrocarbons account for the majority of India's energy use. Together, coal and oil represent about two-thirds of total energy use. Natural gas now accounts for a seven percent share, which is expected to grow with the discovery of new gas deposits. Combustible renewables and waste constitute about one fourth of Indian energy use. This share includes traditional biomass sources such as firewood and dengue, which are used by more than 800 million Indian households for cooking. Other renewables such as wind, geothermal, solar and hydroelectricity represent a 2 % share of the Indian fuel mix. Nuclear holds a one % share.

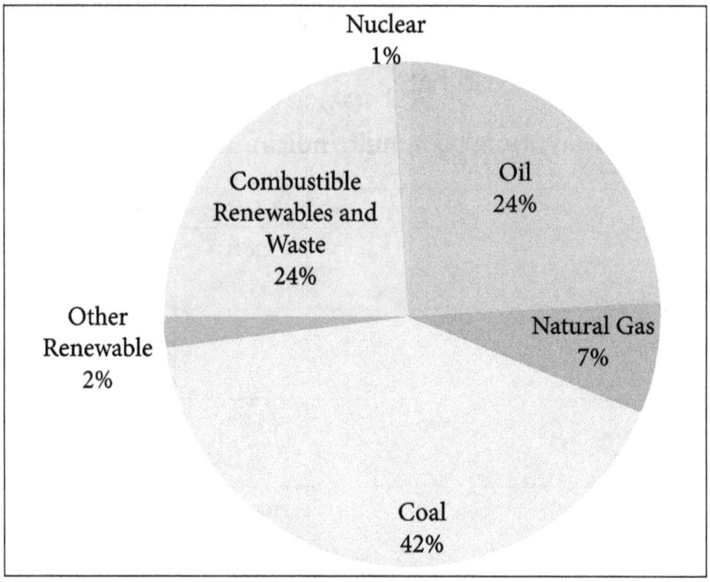

Total Energy Consumption in India by Type (2009).

In 2009, India was the fourth largest energy consumer in the world, after the United States, China and Russia. Despite a slowing global economy, India's energy demand continues to rise. As vehicle ownership expands, petroleum demand in the transport sector is expected to grow in the coming years. While India's domestic energy resource base is substantial, the country relies on imports for a considerable amount of its energy use.

IEA data for 2009 indicate that electrification rates for India were 66 % for the country as a whole. Ninety-four percent of the 404 million that do not have access to electricity live in rural areas, where electrification rates are approximately 50%.

7. Electricity and Electricity Shortages

India currently suffers from a major shortage of generation capacity. In 2008, India had approximately 177 gigawatts (GW) of installed electric capacity and generated 761 billion kilowatt hours. Conventional thermal sources produce more than 80 % of India's electricity. Hydroelectricity, nuclear power and other renewable sources account for the remainder. India also imports marginal amounts of electricity from Bhutan and Nepal and has signed an agreement to begin importing power from Bangladesh.

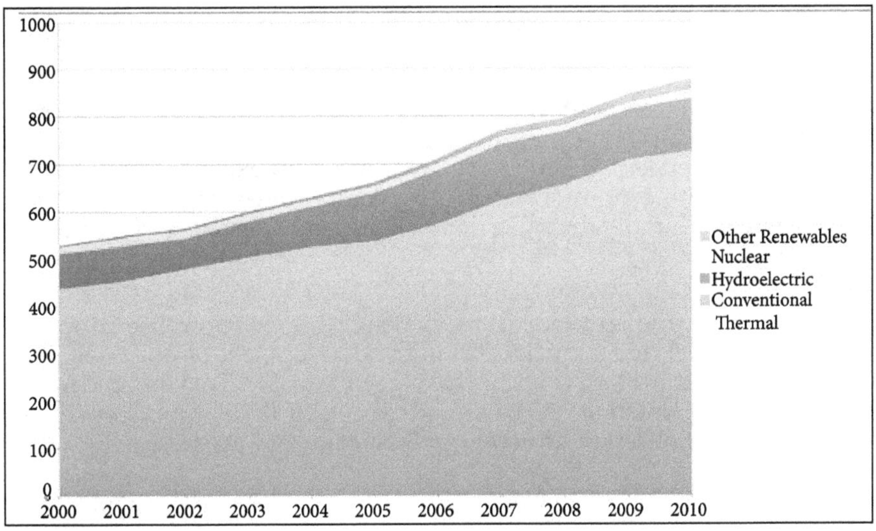

Indian Electricity Generation by Type (billion kilowatt hours), 2000-2010.

India suffers from a severe shortage of electricity generation capacity. According to the World Bank, roughly 40 % of residences in India are without electricity. In addition, blackouts are a common occurrence throughout the country's main cities. Further compounding the situation is that total demand for electricity in the country continues to rise and is outpacing increases in capacity. Additional capacity has failed to materialize in India in light of market regulations, insufficient investment in the sector and difficulty in obtaining environmental approval and funding for hydropower projects. In addition, coal shortages are further straining power generation capabilities. In order to address this shortfall, the Indian government continues to work towards adding capacity. In the IEO2011, EIA projects that electricity consumption in India will grow at an average rate of 3.3 % per year through 2035. To meet this growth, India will have to expand their current generation capacity by 234GW (Energy Information Administration, 2011).

Conclusion

Jawaharlal Nehru's famous description of Bangalore as the city of the future, unburdened by traces of the past has itself become something of a burden for the city to bear. The government's ability to expand and improve services is severely tested given the city's rapid economic and population growth. New opportunities and challenges are coexisting in the growth puzzle of Bangalore. Given the economic expansion and success of Bangalore, it is possible with proper investment and planning that the city could become a model for sustainable growth in the developing world.

Permissions

All chapters in this book are published with permission under the Creative Commons Attribution Share Alike License or equivalent. Every chapter published in this book has been scrutinized by our experts. Their significance has been extensively debated. The topics covered herein carry significant information for a comprehensive understanding. They may even be implemented as practical applications or may be referred to as a beginning point for further studies.

We would like to thank the editorial team for lending their expertise to make the book truly unique. They have played a crucial role in the development of this book. Without their invaluable contributions this book wouldn't have been possible. They have made vital efforts to compile up to date information on the varied aspects of this subject to make this book a valuable addition to the collection of many professionals and students.

This book was conceptualized with the vision of imparting up-to-date and integrated information in this field. To ensure the same, a matchless editorial board was set up. Every individual on the board went through rigorous rounds of assessment to prove their worth. After which they invested a large part of their time researching and compiling the most relevant data for our readers.

The editorial board has been involved in producing this book since its inception. They have spent rigorous hours researching and exploring the diverse topics which have resulted in the successful publishing of this book. They have passed on their knowledge of decades through this book. To expedite this challenging task, the publisher supported the team at every step. A small team of assistant editors was also appointed to further simplify the editing procedure and attain best results for the readers.

Apart from the editorial board, the designing team has also invested a significant amount of their time in understanding the subject and creating the most relevant covers. They scrutinized every image to scout for the most suitable representation of the subject and create an appropriate cover for the book.

The publishing team has been an ardent support to the editorial, designing and production team. Their endless efforts to recruit the best for this project, has resulted in the accomplishment of this book. They are a veteran in the field of academics and their pool of knowledge is as vast as their experience in printing. Their expertise and guidance has proved useful at every step. Their uncompromising quality standards have made this book an exceptional effort. Their encouragement from time to time has been an inspiration for everyone.

The publisher and the editorial board hope that this book will prove to be a valuable piece of knowledge for students, practitioners and scholars across the globe.

Index

Printed in the USA
CPSIA information can be obtained
at www.ICGtesting.com
JSHW052311061123
51544JS00005B/13

9 781641 167550